# 2009 Annual Meeting of the American Electrophoresis Society



Topical Conference at the 2009 AIChE Annual Meeting

## Nashville, Tennessee, USA
## 8-13 November 2009

ISBN: 978-1-61567-914-0

**Printed from e-media with permission by:**

Curran Associates, Inc.
57 Morehouse Lane
Red Hook, NY 12571

**Some format issues inherent in the e-media version may also appear in this print version.**



Printed by Curran Associates, Inc. (2010)

For permission requests, please contact AIChE
at the address below.

AIChE
3 Park Avenue
New York, NY 10016-5991

Phone: (203) 702-7660
Fax: (203) 775-5177

www.aiche.org

**Additional copies of this publication are available from:**

Curran Associates, Inc.
57 Morehouse Lane
Red Hook, NY 12571 USA
Phone: 845-758-0400
Fax: 845-758-2634
Email: curran@proceedings.com
Web: www.proceedings.com

# TABLE OF CONTENTS

# Asymmetric Flows Over Symmetric Surfaces: Capacitive Coupling in Induced Charge Electro-Osmosis

**Todd M. Squires**, Chemical Engineering, University of California at Santa Barbara, Santa Barbara, CA

Tobias S. Mansuripur, Physics, University of California, Santa Barbara, Santa Barbara, CA

Andrew J. Pascall, Chemical Engineering, University of California, Santa Barbara, Santa Barbara, CA

We report curious asymmetric induced-charge electro-osmotic (ICEO) flows over a symmetric, planar gate electrode under applied AC electric fields, whereas symmetric, counter-rotating rolls are expected. Furthermore, the asymmetric component of the flow is consistently directed towards the grounded electrode. We propose that capacitive coupling of the gate electrode to the microscope stage--a comparatively large equipotential surface that acts effectively as a ground--is responsible for this symmetry breaking. This stray capacitance drives the formation of a double layer whose zeta potential is proportional to the potential drop from the electrolyte directly above the gate electrode to the external stage. Therefore, the charge in this ``stray" double layer varies in phase with the driving field, resulting in a rectified, steady flow as with standard ICEO. We experimentally vary the stray capacitance, the electric potential of the stage, and the location of the gate electrode, and find that the effect on the stray flow is qualitatively consistent with the predictions of the proposed mechanism. In the process, we demonstrate that capacitive coupling offers an additional means of manipulating fluid flow over a polarizable surface.

See more of this Session: Advances in Electrokinetics and Electrophoresis - Fundamentals
See more of this Group/Topical: Topical 3: 2009 Annual Meeting of the American Electrophoresis Society (AES)

# Effect of Material Morphology On Electrokinetic-Based Bioseparations: Comparison Between Computational and Analytical Results

**Jyothirmai J. Simhadri**, Chemical Engineering, Tennessee Technological University, Cookeville, TN

**Jennifer Pascal**, Chemical Engineering, Tennessee Technological University, Cookeville, TN

Mario Oyanader, Chemical Engineering, Tennessee Technological University, Cookeville, TN

Holly A. Stretz, Chemical Engineering, Tennessee Technological University, Cookeville, TN

Pedro Arce, Chemical Engineering, Tennessee Technological University, Cookeville, TN

Research involving polymer gels with embedded nanoparticles of varying properties is quite attractive because of the multitude of potential applications, including separation of biomacromolecules, tissue scaffold growth, as well as the analysis of the efficiency and control of drug delivery. The presence of nanoparticles within gels has the potential to modify not only the gel morphology but also the electrokinetic properties of the gels; therefore, these nanoparticles may influence both the electrophoretic transport as well as electro-osmotic flows.

This project focuses on idealized capillary models whose characteristics are useful domains to mimic the gel morphology in the aforementioned nanocomposite gels. Such domains allow for the use of a nonuniform cross section and electrostatic potential along the capillary walls in order to capture the electrophoretic and electrostatic behaviors between the nanoparticles and the gel. Analyses of the effects of various system parameters such as the channel cross sections, electrostatic potential, and aspect ratio, among others are important aspects to be studied either analytically or computationally as shown by previous efforts. In this presentation we will compare computational results obtained from the analysis of electrophoresis in a diverging channel to those obtained analytically based on the use of electrokinetic hydrodynamics (EKHD) concepts coupled with the spatial averaging approach originally introduced by Slattery, Whitaker, and others.

See more of this Session: Advances in Electrokinetics and Electrophoresis - Fundamentals
See more of this Group/Topical: Topical 3: 2009 Annual Meeting of the American Electrophoresis Society (AES)

# Simultaneous High-Resolution Separation and Concentration of Proteins Using Microscale Pore Limit Electrophoresis

Greg J. Sommer, Biosystems Research and Development, Sandia National Laboratories, Livermore, CA

Anup K. Singh, Biosystems Research and Development, Sandia National Laboratories, Livermore, CA

Anson V. Hatch, Biosystems Research and Development, Sandia National Laboratories, Livermore, CA

Here we present our recent work in photopolymerizing precise and well-controlled polyacrylamide porosity gradients in microchannels for microscale pore limit electrophoresis (μ PLE) of native proteins. Pore limit electrophoresis was introduced in the late 1960's as a slab gel-based separation technique in which a mixture of proteins is fractionated along a polymer gradient of decreasing pore sizes reaching dimensions smaller than the effective molecular diameter of analyte species. In this manner, PLE provides high-resolution fractionation of complex samples based on the spatial dependence of each species' electrophoretic pore limit - the porosity at which a protein's electrophoretic mobility is negligible due to its molecular size. In μ PLE, we aim to advance this separation technique by achieving certain advantages-of-scale associated with miniaturizing pore gradient gels, including faster run times and better reproducibility. The pore gradient gel is also useful for estimating size-exclusion membrane thresholds for a broad range of polymer concentrations and protein sizes simultaneously.

We adapted our previously-developed polymer gradient photopolymerization technique [1] to fabricate on-chip pore gradient gels. Porosity was controlled via distributions of acrylamide monomer and bisacrylamide crosslinker. We demonstrate simultaneous separation and concentration of proteins ranging in molecular weight from 21.5kDa - 144kDa along 5mm μ PLE gels spanning 10%T, 2.6%C - 40%T, 12%C. Results show increasing resolution as proteins approach their effective pore limits. However, continued migration beyond this point shows electrophoretic behavior, including increased dispersion, which strays from theoretical predictions based on the Ferguson relationship. We will discuss potential explanations for the electrokinetic phenomena observed in this non-Ferguson regime. We will also show that μ PLE can be used to concentrate dilute samples by exploiting the stacking phenomenon associated with an analyte's decreasing electrophoretic mobility. Concentration factors > 40,000 are easily achieved with dilute (100pM) samples.

[1] G.J. Sommer, A.K. Singh, A.V. Hatch. "On-Chip Isoelectric Focusing Using Photopolymerized Immobilized pH Gradients" Anal. Chem., 2008, 80, 3327-3333.

See more of this Session: Advances in Electrokinetics and Electrophoresis - Fundamentals

See more of this Group/Topical: Topical 3: 2009 Annual Meeting of the American Electrophoresis Society (AES)

# Transverse Migration of a Polyelectrolyte Driven by a Combination of Electrophoresis and Pressure-Driven Flow

**Jason E. Butler**, Chemical Engineering, University of Florida, Gainesville, FL

Rahul Kekre, Department of Chemical Engineering, University of Florida, Gainesville, FL

O. Berk Usta, Chemical and Petroleum Engineering, University of Pittsburgh, Pittsburgh, PA

Anthony J.C. Ladd, Department of Chemical Engineering, University of Florida, Gainesville, FL

The cross-stream migration of a polyelectrolyte flowing within a confining channel under combinations of an electrophoretic driving force and pressure-driven flow is examined. When acting in conjunction, the polymer migrates to the center of the channel and when the pressure gradient and external force act in opposite directions, the polymer can migrate towards the boundaries. Producing this behavior, which agrees qualitatively with published experiments on DNA, requires including hydrodynamic interactions with only partial screening due to the action of the electric field on the counterions surrounding the polyelectrolyte; models of DNA electrophoresis that utilize an assumption of complete screening of the hydrodynamic interactions fail to predict the migration observed in the experiments. The results of the analysis suggest that models of DNA electrophoresis can be improved by including the screened hydrodynamic interactions and comparing with detailed observations from similar experiments.

See more of this Session: Advances in Electrokinetics and Electrophoresis - Fundamentals
See more of this Group/Topical: Topical 3: 2009 Annual Meeting of the American Electrophoresis Society (AES)

# Electrical Conductivity of Aqueous Solutions of Surfactants with Added Electrolytes

**Ezinwa O. Elele**, Chemical, Biological & Pharmaceutical Engineering, New Jersey Institute of Technology, Newark, NJ

Boris Khusid, Chemical, Biological & Pharmaceutical Engineering, New Jersey Institute of Technology, Newark, NJ

Yueyang Shen, Chemical, Biological & Pharmaceutical Engineering, New Jersey Institute of Technology, Newark, NJ

We analyze the electrical conductivity of aqueous solutions of sodium dodecyl sulphate (SDS) and lithium dodecyl sulfate (LDS) with added electrolytes as a function of the surfactant and electrolyte concentrations. A molecular thermodynamic model of surfactant aggregation in aqueous solutions was used to predict the micelle concentration, the average size and shape of aggregates, and the detailed size distribution of aggregates. Based on these characteristics, the electrical conductivity of a solution was then computed within the framework of the classical Onsager theory for a multicomponent electrolyte. Theoretical predictions were compared with experimental data. Measurements were carried out on a broadband dielectric spectrometer BDS-80, Novocontrol. Experiments cover a broad range of the surfactant and electrolyte concentrations, from above the critical micelle concentration (CMC) to infinite dilution. Deviation of theoretical predictions from experimental data provides insight into the formation of micellar aggregates.

This work was supported by NSF Engineering Research Center on Structured Organic Particulate Systems.

See more of this Session: Advances in Electrokinetics and Electrophoresis - Fundamentals
See more of this Group/Topical: Topical 3: 2009 Annual Meeting of the American Electrophoresis Society (AES)

# Effect of Ionic Liquid On Microfluidic Electrophoretic Separation of Phenolic Acids

**Suk Tai Chang**, Sandia National Laboratories, Livermore, CA

Rajiv Bharadwaj, Sandia National Laboratories, Livermore, CA

Seema Singh, Sandia National Laboratories, Livermore, CA

Anup K. Singh, Sandia National Laboratories, Livermore, CA

There is a growing interest in the production of biofuels from a variety of cellulosic feedstock crops such as Switchgrass, Miscanthus, and Corn Stover. The presence of lignin in the plant cell walls is a major source of biomass recalcitrance. Recently, several studies have shown that ionic liquids (ILs) can effectively de-polymerize biomass and enable enhanced enzymatic hydrolysis rates. However, due to a lack of IL-compatible analytical techniques, the molecular level understanding of the biomass deconstruction process is difficult. This limits the rational optimization of biomass pretreatment processes. Our aim is to develop a label-free IL-compatible electrophoresis approach to quantify lignin degradation products. Our approach leverages the autofluorescene of phenolic compounds, such as Ferulic acid and p-Coumaric acid, for label-free detection. We have also investigated the effect of room temperature ionic liquid, 1-ethyl-3-methylimmidazolium acetate (EMIM-Acetate), on electrokinetic properties of glass microchips. Interestingly, even as little as 0.2% (v/v) IL solution can reduce the electroosmotic flow (EOF) by 10-fold at highly basic pH (> 9.0). In addition, there is a distinct timescale associated with the rate of decrease of EOF at a fixed IL concentration. This indicates that EMIM-Acetate adsorbs to the surface of the glass microchip. We are developing electrokinetic models to investigate the dynamics of IL adsorption and EOF suppression in glass microchips. We have demonstrated the microfluidic electrophoretic separation of model phenolic compounds in presence of IL within 60 s with a detection limit of around 5 μg/mL. We are currently extending the approach to analyze real-world cellulosic biomass samples to optimize pretreatment processes for bioenergy production.

See more of this Session: Advances in Electrokinetics and Electrophoresis - Fundamentals
See more of this Group/Topical: Topical 3: 2009 Annual Meeting of the American Electrophoresis Society (AES)

# Electrofocusing Small Organic Ions

**Cornelius F. Ivory**, School of Chemical and Biological Engineering, Washington State University, Pullman, WA

Jeffrey Burke, School of Chemical Engineering and Bioengineering, Washington State University, Pullman, WA

Colin Smith, School of Chemical Engineering and Bioengineering, Washington State University, Pullman, WA

Traditional electrofocusing methods, i.e., isoelectric focusing, are very difficult to use with ionic species that don't have a defined isoelectric point, for example, pharmaceuticals. However, these low molecular weight ions can be separated and concentrated using alternative focusing methods like dynamic field-gradient focusing (DFGF). The problem that arises with small molecules is that DFGF traditionally uses a dialysis membrane to segregate the separation channel from the electrodes and so ions with molecular weights less than about 1000 will leak through membranes with molecular weight cutoffs as low as 100.

This paper describes our efforts to design a column and develop protocols suitable for electrofocusing small, non-amphoteric ions by eliminating the membrane and installing the electrodes directly into the separation channel. Doing this causes three major problems: (1) gas evolution at the electrodes, (2) pH changes adjacent to the electrodes, and (3) solute oxidation-reduction reactions on the electrode surfaces. A combination of experiment and simulation was used to determine whether these problems could be eliminated by changing the design of the separation channel and the protocols used to electrofocus low molecular weight solutes.

See more of this Session: Advances in Electrokinetics and Electrophoresis - Fundamentals
See more of this Group/Topical: Topical 3: 2009 Annual Meeting of the American Electrophoresis Society (AES)

# CONCENTRATION AND SEPARATION OF DNA EMPLOYING INSULATOR-BASED DIELECTROPHORESIS AND DC ELECTRIC FIELDS

*Roberto C. Gallo-Villanueva,[1] Carlos E. Rodríguez-López,[1] Rocío I. Diaz-de-la-Garza,[1] Claudia Reyes-Betanzo,[2] Blanca H. Lapizco-Encinas[1,]**

*[1]Tecnológico de Monterrey, Monterrey, NL, México. [2]Instituto Nacional de Astrofísica, Óptica y Electrónica, Santa María Tonanzintla, Puebla, México*

## Abstract

Dielectrophoresis (DEP) is an electrokinetic technique where particles move due to polarization effects when exposed to non-uniform electric fields, which can be induced either by direct current (DC) or alternating current (AC). This novel technique has a great potential for miniaturization, and it has been successfully applied for the manipulation of a wide array of bioparticles, from biomolecules to microorganisms. Insulator-based dielectrophoresis (iDEP), a method where non-uniform electric fields are created with arrays of insulating structures, has advantages over the traditional electrode-based DEP such as inexpensive fabrication processes and the ability of employing electroosmotic flow to pump liquid when direct current (DC) electric fields are applied; making miniaturization more viable.

In this work, iDEP is used to immobilize and concentrate linear DNA (pET28b) particles inside a microchannel with cylindrical insulating structures fabricated from glass. Results showed successful manipulation of DNA particles with a combination of electroosmotic, electrophoretic and dielectrophoretic forces. DNA particles were observed to be immobilized due to negative dielectrophoretic trapping under applied fields between 500 and 1500 V/cm, in suspending mediums with conductivity of 104 μS/cm and pH of 11.15. Finally, concentration factors varying from 8 up to 24 times the feed concentration were measured employing fields of 2000 V/cm after concentration time-periods of 20 and 40 seconds. The results presented here demonstrate the potential of iDEP for DNA concentration, and open the possibility for fast DNA manipulation for laboratory and large-scale applications.

## Introduction

There is a growing interest in developing separation techniques amenable for miniaturization, since working on the microscale offers attractive advantages over traditional bench scale processes, such as shorter analysis time, reduced sample and reagent consumption, enhanced sensitivity, high resolution, and portability.[1] Microfluidic technology has advanced significantly, since it began almost two decades ago, and has beneficiated numerous analytical applications, from environmental analysis, biomedical testing, food safety to drug development. [2-4] Electrokinetic techniques, such as electrophoresis and dielectrophoresis are among the most employed methods in micro Total Analytical Systems (μTAS).[5]

Dielectrophoresis (DEP) is an electrokinetic transport mechanism experienced by particles due to polarization effects in the presence of non-uniform electric fields; it can be employed with direct current (DC) and alternating current (AC) electric fields. DEP can achieve concentration or separation on a single step, and has been successfully employed to handle a wide array of bioparticles.[6-8] Separation, concentration, and purification of DNA using DEP-based microdevices could become an alternative to a) the traditional gel electrophoresis techniques, which are time consuming, not convenient for very long DNA molecules, and generate significant biohazard waste; and b) chromatographic purification, which have been the traditional approach employed to eliminate contaminants from DNA for large-scale systems.[9-12]. Moreover, the ability to separate nucleic acids with DEP while observing the process could allow the monitoring of their interactions with other molecules (*e.g.* proteins). This opens the possibility for several biotechnological analytical applications, such as on-chip hybridizations and synthesis, gene mapping and genotyping, etc. In addition, the use of electroosmotic flow as the fluid driving force, can prevent DNA degradation caused by shear-stress generated by pressure-driven flow.[13] DEP can be envisioned as technique that could answer many of the requirements for purification of pharmaceutical-grade nucleic acids. Gene therapy and DNA vaccines will become a commercial reality in the following years,[14-16] in fact, large-scale production and purification of nucleic acids is currently on demand for both laboratory and clinical trials.[17,18] Microdevices used in massively parallel systems could accomplish large-scale DNA purification, leading to faster separations while decreasing reagent consumption; making the purification processes more economically favorable.

## Theory

First proposed by Pohl in 1951, dielectrophoresis is particle movement due to polarization effects under the presence of a non-uniform electric field; it can occur with both, alternating current (AC) or direct current (DC) fields.[19] The dielectrophoretic force exerted on a spherical particle can be defined as:

$$F_{DEP} = 2\pi\varepsilon_m r_p^{\,3}\,\mathrm{Re}(f_{CM})\,\nabla E^2 \qquad (1)$$

where $\varepsilon_m$ is the permittivity of the suspending medium, $r_p$ is the radius of the particle, $E$ is the local electric field, and $\mathrm{Re}(f_{CM})$ is the real part of the Clausius-Mossotti (CM) factor expressed as:

$$f_{CM} = \frac{\varepsilon_p^* - \varepsilon_m^*}{\varepsilon_p^* + 2\varepsilon_m^*} \qquad (2)$$

where $\varepsilon_p^*$ and $\varepsilon_m^*$ are the complex permittivities of the particle and the medium, respectively. The complex permittivity is related to the real permittivity $\varepsilon$ by the expression $\varepsilon^* = \varepsilon - (j\sigma/\omega)$, where $\sigma$ and $\omega$ are the conductivity and angular frequency of the applied electric field respectively; while $j=\sqrt{-1}$. The CM accounts for the particle polarizability, a positive value means that the particle is more polarizable than the suspending medium and will be attracted to the regions of higher field intensity. A negative value of CM, means that particles are less polarizable than the medium and will be repelled from the regions of higher field intensity. When a DC electric field is used, the CM factor can be expressed in terms of the real conductivities of the particle ($\sigma_p$) and the suspending medium ($\sigma_m$): [20]

$$f_{CM} = \frac{\sigma_p - \sigma_m}{\sigma_p + 2\sigma_m} \qquad (3)$$

When DC electric fields are employed with iDEP, electrophoresis (EP) and electroosmotic flow (EOF) are also present. Electrokinetics (EK) is the superposition of EP and EOF. For a microchannel with a negative surface charge, such as glass, the electrokinetic velocity ($v_{EK}$,) can be defined as follows, considering the net direction of the flow as positive:

$$v_{EK} = \mu_{EK} E = (\mu_{EO} - \mu_{EP})E \quad (4)$$

$$v_{EO} = \mu_{EO} E \quad (5)$$

$$v_{EP} = \mu_{EP} E \quad (6)$$

where $\mu_{EO}$ and $\mu_{EP}$ are the electroosmotic and electrophoretic mobilities, and $v_{EO}$ and $v_{EP}$ are the electroosmotic and electrophoretic velocities, respectively. In order to achieve trapping of particles, the dielectrophoretic velocity must overcome that of electrokinetics.[21] This is possible only under high electric fields, since the former $v_{EK}$ is linearly related to the field, while $v_{DEP}$ has a relationship of second order with the applied field:

$$v_{DEP} = -\mu_{DEP} \nabla E^2 \quad (7)$$

where $v_{DEP}$ and $\mu_{DEP}$ are the dielectrophoretic velocity and mobility, respectively.

## Materials and Methods

### Microdevice

The experiments were carried out employing a microdevice made from two D263 borosilicate glass wafers. The top wafer contains drilled reservoirs while the bottom glass wafer contains microchannels that were patterned employing standard wet etching techniques. The microchannel used for experimentation is 10.16-mm-long, 2-mm-wide, and 7-µm-deep; and contains an array of cylindrical posts straddled by an inlet and outlet reservoirs. The posts had a diameter of 470 µm, arranged 510 µm center-to-center as shown in Figure 1. The post array had a total of 32 cylindrical posts, arranged in eight columns of four posts each. The first row of posts on either side has a "dove tail" geometry to prevent particles from colliding against the posts and plugging the system.

### DNA sample preparation

*Escherichia coli* TOP10F' cells harboring the pET28b(+) (Novagen Inc., Madison, WI, USA) vector (5,369 base pairs) were cultured overnight at 37°C in Luria-Bertani (LB) media supplemented with kanamycin (50 $\mu$g/ml). Plasmidic DNA was extracted from bacteria using PureLink™ HiPure Plasmid Midiprep Kit from Invitrogen (Cat. No. K2100-05, Carlsbad, CA, USA). All purified plasmid DNA was diluted in ultrapure distilled water, free of DNases and RNases (Cat. No. 10977-015, Invitrogen, Carlsbad, CA, USA). Linearization of plasmid DNA was achieved by selective cleavage using HindIII restriction endonuclease (New England Biolabs, Ipswich, MA, USA). Linear DNA was then precipitated by adding 10% and 250% of the solution volume of $CH_3COONa$ 3M and ice-cold ethanol 95%, respectively,followed by centrifugation (14,000 g for 20 minutes).[22] The supernatant was discarded and the precipitated DNA was washed with 70% ethanol and centrifuged a second time (14,000 g for 10 minutes). The dried pellet was resuspended in water to the desired concentration, which was measured by absorbance at 260 nm, and general purity assessed by an UV-scan from 220 to 320 nm using NanoDropTM1000 spectrophotometer (Thermo Scientific, Waltham, MA, USA). Fluorescent staining of DNA was done with Syto 9, excitation/emission 483/503 nm (Invitrogen, Carlsbad, CA, USA).

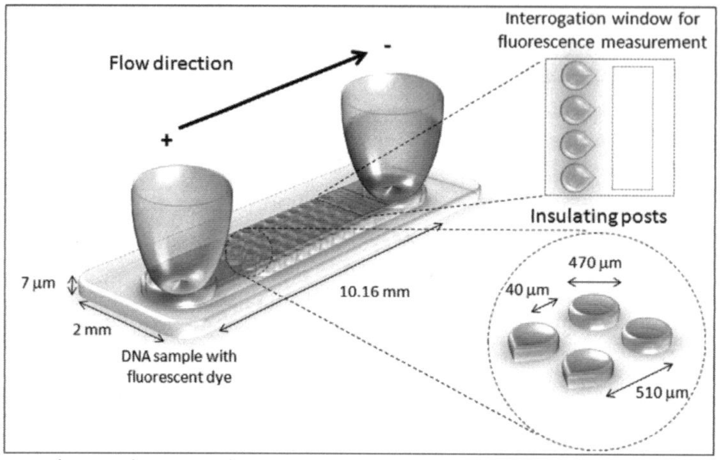


**Figure 1.** Experimental set-up showing the schematic representation of the microchannel employed, where DNA samples are placed at the inlet reservoir and an electrical field is applied between the two reservoirs.

**Equipment and suspending medium**

Direct current electric fields were applied using a 3000 V high voltage sequencer, model HVS448 (LabSmith, Livermore, CA, USA) by employing platinum-wire electrodes. Dielectrophoretic behavior of DNA particles was recorded in the form of video and pictures by utilizing an inverted epifluorescence video microscope for microfluidics, model SVM340 (LabSmith, Livermore, CA, USA). A 4X microscope objective was employed for all experiments. Suspending mediums were prepared from bidistilled water with the use of Tween-20 and triethanolamine (TEA) as DNA stabilizers reaching concentrations of 1.5 % and 1.5 M, respectively, and conductivity and pH were measured and adjusted to desired values in the following ranges: 100 to 120 µS/cm for conductivity and 10.8 to 11.15 for pH by adding NaOH and $KH_2PO_4$.

**Concentration factor estimations**

DNA concentration factor of the sample employing iDEP was evaluated by performing fluorescence measurements in an interrogation window located just after the outlet of the post array as shown in Figure 1. A set of experiments was performed, following the next steps: 1) Electroosmotic flow was generated by applying a low voltage of 200 V/cm for five seconds to record and measure the initial fluorescence of the moving sample. 2) Then the applied field was increased to 2000 V/cm for 20 or 40 seconds in order to dielectrophoretically trap DNA particles. 3) Finally, the field was lowered back to 200 V/cm, in order to release the concentrated particles from the dielectrophoretic traps, by means of EOF; and pushing the plug of particles through the interrogation window resulting in an increase of the fluorescence signal. Concentration factor (CF) is determined by dividing the maximum fluorescence value of the plug of concentrated sample (CP) by the fluorescence signal of the sample before dielectrophoretic trapping (CB):

$$CF = \frac{CP}{CB} \qquad (8)$$

## Results and discussion

### Dielectrophoretic trapping of DNA particles

In accordance to the theory, under low applied fields, when DEP is negligible, EOF will dominate the system and sample will move from the positive to the negative electrode.[21,23] As the electric field is increased, weak trapping will occur as DEP and EK forces struggle to overcome each other. When the applied field is sufficiently strong, DEP force will be great enough to overcome EK, and then DNA particles will be negatively dielectrophoretically immobilized, forming a band of trapped particles prior to the regions of higher field intensity (narrow regions between the cylindrical posts) until the field is lowered to release the particles from the traps. Figure 2 shows a sequence of images depicting the behavior described above, these results were obtained employing a suspending medium with $\sigma_m$=104 μS/cm and pH of 11.15. Figure 2a shows DNA particles flowing with EOF under a field of 500 V/cm, where some dielectrophoretic effect is observed at the constriction. Negative dielectrophoretic trapping of DNA particles was clearly observed by increasing the applied field to 1000 V/cm (Fig. 2b). Strong DNA trapping is achieved when the field is increased to 1500 V/cm (Fig. 2c), observing even stronger "waves" of repelled particles. As one can see from these results, successful DNA immobilization due to negative dielectrophoretic trapping is obtained, that increases with the applied field. Due to the small size of DNA particles, high applied fields are required, as DEP force is directly proportional to $r_p^3$, these high fields can cause DNA degradation. Agarose gel electrophoresis analyses were performed with samples extracted from the inlet and outlet reservoirs of the microchannel. Figure 2d presents the results of these analyses, showing that denaturalization and/or degradation of the sample is negligible under the strong electric fields employed.

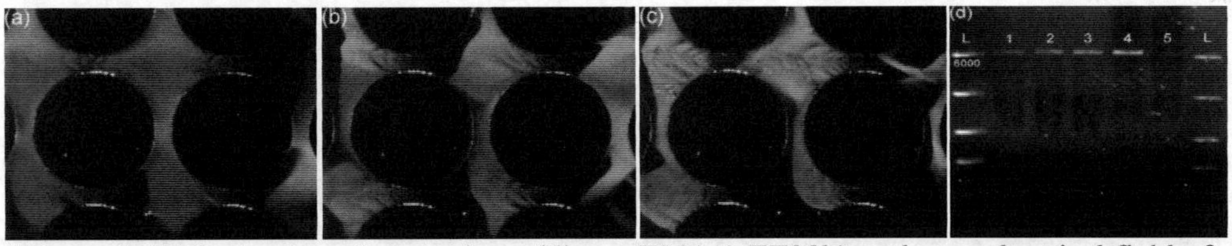


**Figure 2.** Dielectrophoretic trapping of linear DNA (pET28b) under an electrical field of: (a) 500 V/cm, (b) 1,000 V/cm, and (c) 1,500 V/cm, employing a suspending medium with pH =11.15 and $\sigma_m$ = 104 μS/cm. (d) Agarose gel electrophoresis of linear DNA particles as follows: ladder (L), DNA stock (1), dyed DNA stock (2), inlet reservoir (3), sample recovered from the post array (4), sample recovered from outlet reservoir (5). Well defined bands at the same molecular weight (5.3 Kbp) show that DNA was not broken or denatured by dielectrophoretic trapping (molecular weight of the ladder is shown left).

### Determination of concentration factors of DNA particles

Process efficiency can be determined in terms of a concentration factor, which relates the concentration of DNA particles reached after DEP trapping. As mentioned in the experimental section, DNA concentration was measured employing fluorescence signals. Figure 3 shows the fluorescence measurements obtained by following the procedure described in the experimental section. With the field of 200 V/cm,the average fluorescence obtained was 1.5 (arbitrary units of fluorescence), which should correspond to a DNA concentration of 50 μg/mL. Then, the field was increased to 2000 V/cm, employing two different processing times of 20 or 40 seconds, after which it was lowered back to 200 V/cm to release the plug of concentrated particles achieved by dielectrophoretic trapping and measure the increase in fluorescence. Maximum fluorescence values of 12.31 and 36.69 (arbitrary units) were obtained, which correspond to CF's by employing Equation (8) of 8.20 and 24.26 for concentration times of 20 and 40 seconds, respectively, more than an order of magnitude increase compared to feed concentration. This means that the DNA released from the posts was from 8 up to 24 times more concentrated than the original sample. Considering that the feed concentration was 50 μg/mL, the maximum peak concentration obtained was around 1,200 μg/mL. These CF values were obtained employing a suspending medium with a conductivity of 104 μS/cm and pH of 10.8. According to Du *et al.*,[24], there is a nonlinear relation between DNA concentration and fluorescence, *i.e.*, the fluorescence signal obtained does not increases linearly with DNA concentration; fluorescence increases to lesser extent. Therefore, the concentration factor measured with fluorescence is lower than the real concentration factor obtained [24]; this means that our experiments could have produced even higher concentration factors than the values we are reporting. It is important to note that DNA particle concentration was achieved with processing times of around two minutes (Figure 3). This fast concentration process represents a significant advantage over traditional laboratory techniques for DNA concentration such as DNA precipitation and lyophilization.

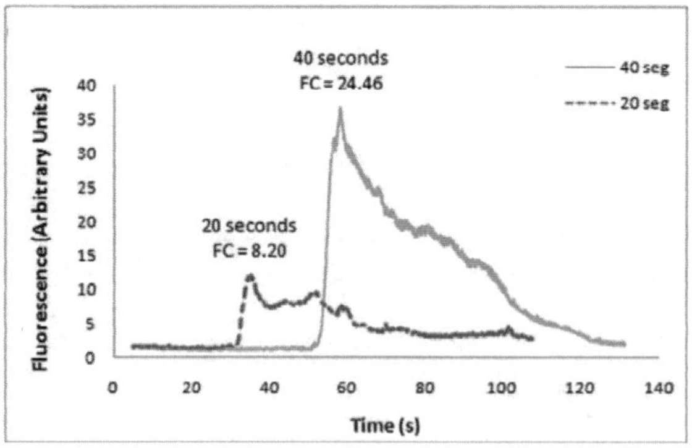


**Figure 3.** Fluorescence measurements as a function of time obtained for estimation of concentration factor of DNA particles.

## Conclusions

This research work presents the manipulation and concentration of linear DNA particles (pET28b) with insulator-based dielectrophoresis (iDEP).Direct current electric fields were employed allowing for electroosmotic flow to pump the flow across the microchannel. Experimental results showed that by varying the magnitude of the applied field between 500 and 1500 V/cm it was possible to control the degree of negative dielectrophoretic trapping, from DNA particles under weak trapping or particles exhibiting strong trapping, suitable to achieve significant sample concentration. These results demonstrated that balancing electrophoretic, electroosmotic and dielectrophoretic forces allows for control of DNA particle manipulation by adjusting operating conditions, such as the magnitude of the applied direct current electric field. In order to test the potential of this system for concentration of DNA, an 8 and 24-fold concentration were achieved by applying a field of 2000 V/cm for processing times of 20 and 40 seconds, respectively.

Insulator-based dielectrophoresis is still a nascent technique, and its application has just started to grow. The results obtained from this study demonstrate that simple iDEP systems, such a microchannel with cylindrical posts, can be employed to manipulate small DNA particles using a combination of electroosmotic, electrophoretic and dielectrophoretic forces. Concentration of DNA particles, at a different level of dielectrophoretic trapping, with our basic system was demonstrated. Concentration factors of DNA samples were also estimated, confirming the great potential of this technique for bioparticle manipulation and concentration, moreover for nano-bioparticles, such as DNA. It is expected that further development of iDEP will make a significant impact on many biotechnological fields, such as gene therapy and DNA vaccines.

## Acknowledgments

The authors would like to acknowledge the financial support provided by the grant CONACYT-CB-2006-53603. The authors are grateful for the financial support provided by *Cátedras de Investigación* (CAT122 and CAT142) of *Tecnológico de Monterrey.*

## References

1 Whitesides G.M., (2006), "The origins and the future of microfluidics," *Nature* 442(7101), pp.368-373.

2 Gardeniers H., Van Den Berg A., (2004), "Micro- and nanofluidic devices for environmental and biomedical applications.," *Int. J. Environ. Anal. Chem.* 84(11), pp.809-819.

3 Escarpa A., González M.C., López-Gil M.A., Crevillén A.G., Hervás M., García M., (2008), "Microchips for CE: Breakthroughs in real-world food analysis," *Electrophoresis* 29(24), pp.4852-4861.

4 Kang L., Chung B.G., Langer R., Khademhosseini A., (2008), "Microfluidics for drug discovery and development: From target selection to product lifecycle management," *Drug Discov. Today* 13(1-2), pp.1-13.

5 Gonzalez C.F., Remcho V.T., (2005), "Harnessing dielectric forces for separation of cells, fine particles and macromolecules.," *J. Chromatogr. A* 1079(1-2), pp.59-68.

6 Voldman J., (2006), "Electrical forces for microscale cell manipulation," *Ann. Rev. Biomed. Eng.* 8, pp.425-454.

7 Lapizco-Encinas B.H., Rito-Palomares M., (2007), "Dielectrophoresis for the manipulation of nanobioparticles," *Electrophoresis* 28(24), pp.4521-4538.

8 Gascoyne P.R.C., Vykoukal J., (2002), "Particle separation by dielectrophoresis," *Electrophoresis* 23(13), pp.1973-1983.

9 Prazeres D.M.F., Ferreira G.N.M., Monteiro G.A., Cooney C.L., Cabral J.M.S., (1999), "Large-scale production of pharmaceutical-grade plasmid DNA for gene therapy: problems and bottlenecks," *Trends Biotechnol.* 17(4), pp.169-174.

10 Danquah M.K., Forde G.M., (2008), "Large-volume methacrylate monolith for plasmid purification: Process engineering approach to synthesis and application," *J. Chromatogr. A* 1188(2), pp.227-233.

11 Levy M.S., O'Kennedy R.D., Ayazi-Shamlou P., Dunnill P., (2000), "Biochemical engineering approaches to the challenges of producing pure plasmid DNA," *Trends Biotechnol.* 18(7), pp.296-305.

12 Guerrero-Germán P., Prazeres D., Guzmán R., Montesinos-Cisneros R., Tejeda-Mansir A., (2009), "Purification of plasmid DNA using tangential flow filtration and tandem anion-exchange membrane chromatography," *Bioproc. Biosys.Engg.* 32(5), pp.615-623.

13 Levy M., Collins I., Yim S., Ward J., Titchener-Hooker N., Shamlou P., Dunnill P., (1999), "Effect of shear on plasmid DNA solution," *Bioproc. Bioengg.* 20, pp.7-13.

14 Li S.-D., Huang L., (2006), "Gene therapy progress and prospects: non-viral gene therapy by systemic delivery," *Gene Ther.* 13(18), pp.1313-1319.

15 Rice J., Ottensmeier C.H., Stevenson F.K., (2008), "DNA vaccines: precision tools for activating effective immunity against cancer," *Nat Rev Cancer* 8(2), pp.108-120.

16 Donnelly J.J., Wahren B., Liu M.A., (2005), "DNA Vaccines: Progress and Challenges," *J. Immunol.* 175(2), pp.633-639.

17 Freitas S., Santos J.L., Prazeres D.M.F., (2006), "Plasmid DNA Production," in Biwer A, Heinzle E, Cooney C (Editors), *Development of Sustainable Bioprocesses: Modeling and Assessment*, Wiley, New York (USA), pp. 271-285.

18 Bower D., Prather K., (2009), "Engineering of bacterial strains and vectors for the production of plasmid DNA," *Appl. Microbiol. Biotechnol.* 82(5), pp.805-813.

19 Pohl H.A., (1951), "The Motion and Precipitation of Suspensoids in Divergent Electric Fields," *J. Appl. Phys.* 22(7), pp.869-871.

20 Benguigui L., Lin I.J., (1984), "Phenomenological Aspect of Particle Trapping by Dielectrophoresis," *J. Appl. Phys.* 56, pp.3294-3297.

21 Martínez-López J.I., Moncada-Hernández H., Baylon-Cardiel J.L., Martínez-Chapa S.O., Rito-Palomares M., Lapizco-Encinas B.H., (2009), "Characterization of electrokinetic mobility of microparticles in order to improve dielectrophoretic concentration," *Anal. Bioanal. Chem.* 394(1), pp.293-302.

22 Altermann E., Russell W.M., Azcarate-Peril M.A., Barrangou R., Buck B.L., McAuliffe O., Souther N., Dobson A., Duong T., Callanan M., (2005), "Complete genome sequence of the probiotic lactic acid bacterium Lactobacillus acidophilus NCFM," *Proc. Natl. Acad. Sci. USA* 102(11), pp.3906-3912.

23 Cummings E.B., Singh A.K., (2003), "Dielectrophoresis in Microchips Containing Arrays of Insulating Posts: Theoretical and Experimental Results," *Anal. Chem.* 75(18), pp.4724-4731.

24 Du J.-R., Juang Y.-J., Wu J.-T., Wei H.-H., (2008), "Long-range and superfast trapping of DNA molecules in an ac electrokinetic funnel," *Biomicrofluidics* 2(4), pp.044103.1-14.

## Direct Control of Biological Assembly Using Electrokinetic Forces

*Michael B. Sano, Andrea Rojas, Paul Gatenholm, Rafael V. Davalos*
*Virginia Tech, Blacksburg, VA*

### *Introduction*

Traditional manufacturing methods for small-scale devices, such as microfabrication, have limitations with regard to control of shape and size. Since top-down manufacturing methods are inadequate for manufacturing larger devices with complex nano-sized features, there is an emerging interest in using biological systems for bottom-up manufacturing. Biofabrication, the combination of biology and microfabrication, may be the future solution for the production of complex 3D architectures with nanoscale precision[1]. This process is advantageous since many organisms are preprogrammed to fabricate complex structures. It has been previously demonstrated that bacteria can be magnetically manipulated to create complex magnetite nanoparticle chains[2] or be ultrasonically processed to create hollow metal chalcogenide nanostructures[3], and genetically engineered viruses can be used to fabricate ordered arrays of quantum dots[4]. A vast number of other potentially useful biological processes exist and biological assembly can be affected by various stimuli such as electrical fields, magnetic fields, temperature, pH, or chemical gradients.

Cellulose, a natural polymer produced by the majority of plants, can also be assembled into nanofibrils by bacteria[5]. Cellulose synthesis by *Acetobacter xylinum* is a complex process which involves the polymerization of single glucose molecules into beta-1,4-glucan chains, the extracellular "extrusion" of the linear chains, and the assembly of the cellulose chains into ribbon-like fibrils with typical dimensions on the nanometer scale[6]. As a result of the motion of these bacteria, a three-dimensional network is produced. The water binding ability of these cellulose ribbons, due to their large surface area and high concentration of hydrophilic hydroxyl groups, yields a hydrogel-like material with unique mechanical properties[7]. *A. xylinum* is difficult to culture in traditional fermentation technology[8] and when agitated, the bacteria can switch off the cellulose production[9]. When left in static culture, *A. xylinum* produces a dense, completely randomized 2D cellulose network, as seen in Figure 1, which increases in thickness as the bacteria rise to stay near the oxygen rich liquid-air boundary.

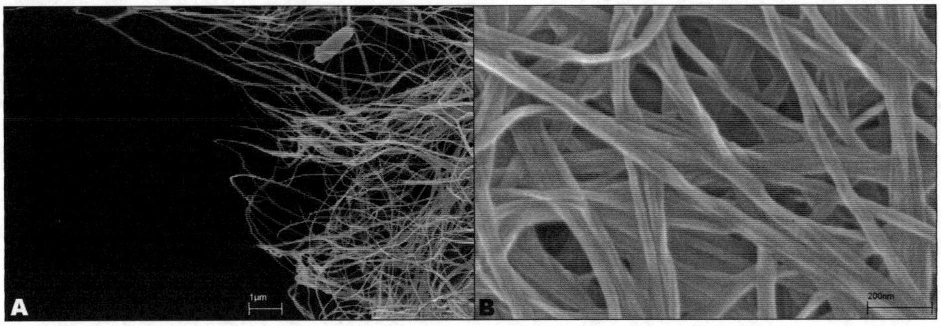


Figure 1: A single A. xylinum cell is entangled in b) the dense random cellulose network it helped produce.

This cellulose network is particularly viable as a material for biomedical applications due to its' high purity, biocompatibility, mechanical integrity, hydroexpansivity, and its stability under a wide range of conditions[10]. Bacterial cellulose has been successfully evaluated in several biomedical applications including blood vessel replacement, meniscus, and bone scaffolds [11-13]. Existing scaffold fabrication techniques suffer from fundamental manufacturing limitations that have, to date, prevented their clinical translation. These limitations result from the distinct lack of processes capable of reproducibly creating structures on the nano-, micro-, and millimeter scales that adequately promote cell growth and function.

We have made steps towards overcoming this limitation by manipulating the shape and morphology of bacterial cellulose material by delivering oxygen to the culture media through a polymeric interface[14]. Other groups such as Putra *et al.* found that *A. xylinum* grown on permeable polydimethylsiloxane (PDMS) substrate showed birefringence indicating some degree of uniaxialfiber alignment [15]. They have also found that bacterial cellulose tubes grown on silicone tubing showed nanofiber orientation along the longitudinal axis [16]. Uraki *et al.* were able to align cellulose nanofibers along honeycomb-patterned microgrooves in an agarose film scaffold[17]. Kondo *et al.* found that bacterial cellulose grown on a molecular substrate of aligned glucan chains resulted in epitaxially aligned bacterial cellulose nanofibers [18]. While these modifications are limited to the fibrils located near the solid-liquid interface, these researchers shed light on the need for controlled orientation of the nanofibers. Here we present a method of direct control over bacterial cellulose nanofiber orientation using electrokinetic forces.

### *Mathematical Methods*

The application of a uniform electric field to an ionic liquid in a microfluidic field gives rise to electrical double layer (EDL) formation along the channel wall. The ions closest to the channel walls are subject to strong electrostatic forces which cannot be overcome by thermal diffusion. As result, these ions are statically bound to the surface of the channel forming a fixed Stern layer. The electrostatic force within the EDL deteriorates further from the channel surface and mobile ions begin to move parallel to the EDL. The net effect of ionic drag caused by the mobile ions on the bulk fluid induces a phenomenon known as *electro-osmotic(EO) flow.*[19] The velocity of an ionic fluid under EO flow ($v_{eo}$) is calculated by

$$v_{eo} = \mu_{eo}\vec{E}$$

where $\mu_{eo}$is the elctro-osmotic mobility of the ionic fluid and $\vec{E}$ is the magnitude of the applied electric field. The electro-osmotic mobility

$$\mu_{eo} = -\frac{\epsilon\zeta}{\eta}$$

is a function of the surface potential between the solid and liquid phase ( $\zeta$) and the viscosity ($\eta$) of the fluid [20].

An EDL will additionally form around a charged particle (or cell) placed in an infinite ionic liquid under a uniform field. In the case of a positively charged cell, a double layer consisting of an excess of positive ions will form. The cell will then be driven towards the region of highest positive potential by a Coulombic force ($\vec{F_c}$) which is proportional to the net charge of the cell (q) and equal to

$$\vec{F_c} = q\vec{E}$$

[21]. This force is also known as *electrophoresis* and the resulting electrophoresis velocity ($\vec{V}_{ep}$) of a spherical cell can be calculated using the Hückel equation[20]

$$\vec{V}_{ep} = \mu_{ep}\vec{E}$$

This velocity is a function of the applied field and the electrophoretic mobility ($\mu_{ep}$) of the cell which is a function of the surface potential between the cell, the surrounding medium, and the permittivity of the surrounding media. The net velocity of the cell as result of these two forces is referred to as the *electrokinetic velocity* ($\vec{V}_{ek}$)

$$\vec{V}_{ek} = (\mu_{eo} + \mu_{ep})\vec{E}$$

where $\mu_{eo}$and $\mu_{ek}$are the electro-osmotic and electrophoretic mobilities respectively. These two terms summed are often referred to as the electrokinetic mobility of a particle ($\mu_{ek}$).

$$\vec{V}_{ek} = \mu_{ek}\vec{E}$$

***Experimental Methods***

Several stages of experimentation were conducted to test electrokinetic control over bacterial cellulose production: cellulose production in test tubes, 100mL beakers, micro-growth chambers, and OptiCell chambers,

***Culture Methods***

The strain *Acetobacter xylinum* subsp.sucrofermentas BPR2001, (700178, American Type Culture Collection) was used for all experiments. Modified fructose media with an addition of corn steep liquid (CSL)[22] was used as the culture media. Six cellulose-forming colonies were cultured for 2 days at 30ºC in a rough flask (nominal volume, 300ml; working volume, 100ml) yielding a cell concentration of 3.7*106 cfu/ml. The bacteria were then liberated by vigorous shaking and inoculated into new culture media.

***Cellulose production environments***

Evaluation of the effects of an electric field on cellulose production was performed in three macro environments, 15mL test tubes, *100mL beakers,* and 10mL Opticell 1100 cell culture chambers and in micro-growth chambers.

*Test Tubes*

2.5mL of inoculated culture media were added to 15mL test tubes. Aluminium electrodes were inserted into two small holes previously drilled into the lids. 1V DC (approximately 0.67V/cm) was then applied across the electrodes and cultivation was allowed to persist for 24 hours before cellulose production was analyzed.

*100mL Beakers*

10mL of inoculated culture media was added to 100mL beakers and aluminium electrodes were inserted along opposing edges of the beaker. Voltages of 10V DC through 20V DC (approximately 2.2V/cm to 4.4V/cm) were applied across the electrodes and cultivation was allowed to persist for 20 minutes.

*OptiCell*

15mL of inoculated culture media was added to Opticell 1100 cell culture chambers (Thermo Fisher Scientific, Waltham, MA). *22Gage* 99.99% aluminum electrodes were then carefully inserted into the chambers via the syringe inlets. Voltages of 0.5V to 5.0V DC (0.077V/cm to 0.77V/cm) were applied across the electrodes and cultivation was allowed to persist for 5 days. The electric field was then removed and the culture chambers were heated in a water bath at 60C for a minimum of 1 hour to kill any A. xylinum cells.

*Fabrication of Micro-Growth Chambers for Scaffold Growth*

Elastomer encased micro-growth chambers were fabricated to evaluate the effects of electromagnetic fields on cellulose production by *A. Xylinum.* Microchambers stamps were fabricated by cleaving unaltered 500 micron thick silicon wafers were into rectangles measuring 4.5x0.5cm. These rectangles were then adheared a glass slide.

A liquid phase elastomer was made by mixing PDMS monomers and the curing agent in the ratio of 10:1 (Sylgrad 184, Dow Corning, USA). The bubbles in the liquid PDMS were removed by exposing the mixture to vacuum for an hour. The PDMS liquid was then poured onto the silicon master which was encased by Aluminium foil, producing a thickness of 1cm. Then 15 minutes was allowed for any remaining bubbles to float to the surface. The PDMS was then cured for 10 minutes at 150 C and peeled off the stamp. Inlet ports to each channel were punched using sharpened dispensing needles (JG14-1.0P, Jensen Global Inc, Santa Barbara, CA).

*PDMS-PDMS Reversible Bond*

The PDMS device layer of the micro-growth chamber and a flat piece of previously prepared PDMS were washed and dried as described above. Both were then exposed to air plasma for 2 minutes on high. After exposure, the layers were immediately brought into contact and pressure was applied as described above. This yielded a semi-permanent, water tight bond and the devices could be separated by gently pulling the two layers apart allowing samples to be removed after culture. Growth chambers remained under pressure a minimum of 10 minutes and were stored in a vacuum chamber until use.

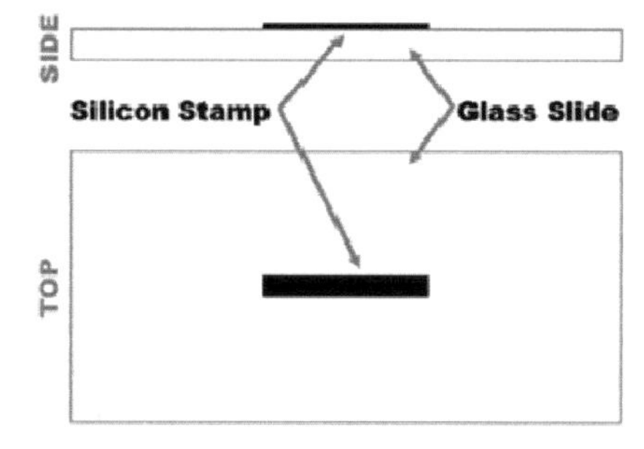


Figure 2: Schematic of rectangular silicon stamp used to create a micro-culture chamber for the production of aligned cellulose nano-fibers.

*Micro-Growth Chamber Cellulose Production*

The micro-growth chamber was first cleaned with ethanol and rinsed with DI water before being primed with 100uL of inoculated culture media. 1000uL pipette tips were inserted into the inlets of the micro-environment and used as media reservoirs. Aluminium electrodes were then placed in the reservoirs and electric fields between 0.1 and 5V/cm were applied. Cultivation was allowed to proceed for 72 hours before the electric field was removed. The micro-environments were then opened and the resulting scaffolds were placed in liquid nitrogen to halt cellulose production.

The scaffolds were then freeze dried in a Labonco FreeZone 2.5 Plus (Labconco Corp., Kansas City, Missouri) freeze dryer for 48 hours without any further processing to leave the bacterial cells *in situ.* 5nm of gold was then deposited on the scaffold and Field Emission Scanning Electron Microscopy (FESEM) was conducted at a working distance of 6mm and 5kV electron beam intensity using a LEO Zeiss 1550 FESEM (Carl Zeiss SMT, Oberkochen, Germany).

### Results and Discussion

Cellulose production in a test tube is altered dramatically under the presence of an electric field. In static culture with no electric field, cellulose production occurs only at the liquid-air boundary. Under the application of an electric field, cellulose production is induced and applied field of 0.67V/cm resulted in significant cellulose production below the liquid-air boundary as shown in Figure 3. Fiber alignment was not visible under these conditions since 3D effects including gravity and buoyancy influenced the network structure.

Figure 3: Cellulose production modified by application of electric field

Experiments in 10mL beakers attempted to minimize 3D effects and to limit the influence of macro-scale forces including gravity and buoyancy. Cellulose production was immediately visible at the anode during the application of 10V and 15V DC. This is likely due to the production of oxygen via hydrolysis. As time progresses a circular region of cellulose around the anode becomes visible. This circular region begins to deform and stretch out towards the cathode as shown in Figure 4. This process is likely due to electrokinetic forces driving the bacterium from one edge of the beaker to the other. The application of 20V results in minimal cellulose production and the cells may be moving too fast to produce cellulose chains or their metabolic processes may be impeded in some manner.

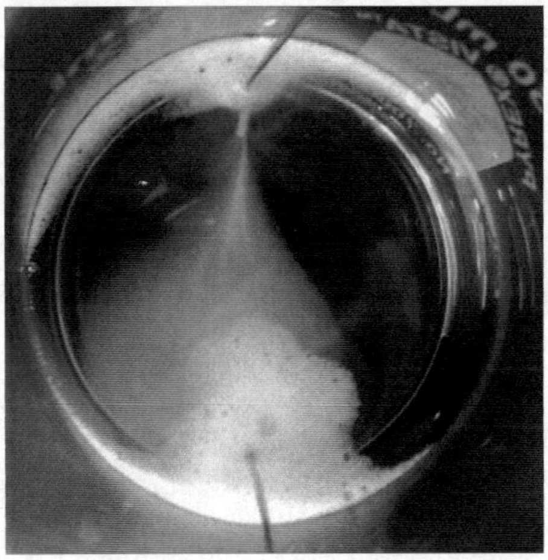

Figure 4: BC Production after 20 minutes in a 100mL beaker w/ 10V applied by aluminum electrodes

The OptiCell culture chambers provided an ideal, confined, 2mm deep chamber to observe cellulose production under the application of an electric field. These experiments were conducted with the OptiCell chambers orientated so that the gravitational force was parallel with the electrodes. Without an applied electric field a random cellulose network was produced between the oxygen permeable

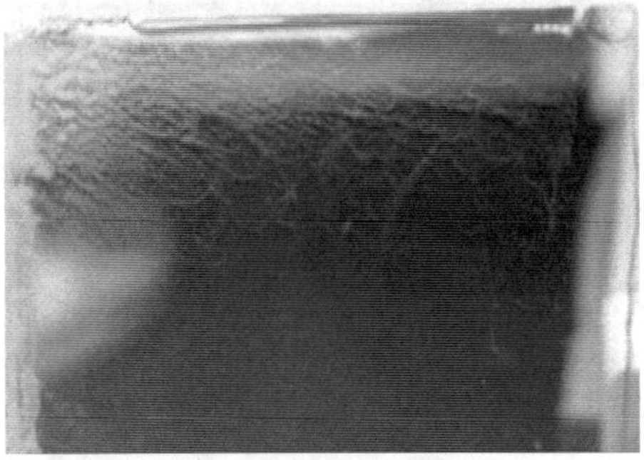

Figure 5: Fibers stretching across OptiCell culture platform produced by the application of 2.0V

membranes. Application of 5V yielded some cellulose; however, cellulose production was notably diminished. For smaller applied fields, such as 0.33V/cm shown in Figure 5, cellulose fibers could be seen extending across the chamber after 24 hours. These fibers are not exactly perpendicular to the electrodes due to gravitational forces.

The micro-growth chambers provided the optimum platform for producing cellulose scaffolds for visualization. Without an applied electric field, the bacteria produce a dense random network as reported previously by numerous authors. Under high electrical fields, the bacteria are moved too quickly, or their metabolic processes are inhibited by the applied field, and cellulose production is switched off. However, there are experimental conditions in which the motion of the bacteria can be controlled while simultaneously producing cellulose. Specifically, electric fields between 0.25V/cm and 1.0V/cm were used to guide the bacteria through the micro-growth chamber by electrokinetic forces. Under these experimental conditions, the bacterial cells are being controlled with velocities on the order of 1 micron/s. Within this range, variations in the strength of the applied field change the morphology of the cellulose structure produced.

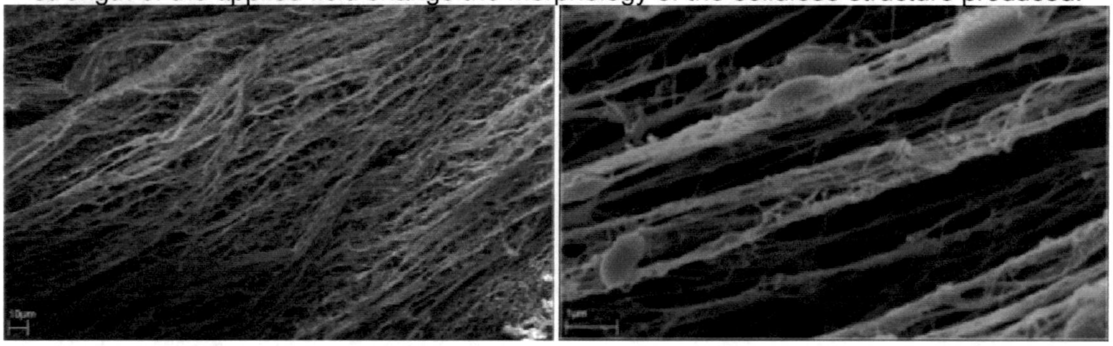

Figure 6: Field Emission Scanning Electron Microscopy (FESEM) of the cellulose network produced by *A. xylinum* in a micro-growth chamber environment under (left) 0.303V/cm and (right) 0.45V/cm electric fields.

We have successfully demonstrated direct control over a biofabrication process to create a structure that is morphologically different from the material produced naturally and Figure 6(left) shows an FESEM image of the cellulose produced under an electric field of 0.303V/cm. The interwoven strands of nanocellulose fibrils are aligned in the direction of the applied electrical fields to which the bacteria were exposed. Increasing the field strength to 0.45V/cm produces a more finely stranded cellulose structure as visible in Figure 6(right). These figures are in clear contrast with Figure 1, which shows a randomly distributed network of nanocellulose. The ellipsoid shaped particles on top of the strands in Figure 6(right) are the bacteria that have been fixed to the cellulose fibers during the freeze drying process. Inspection of the branching nanofiber network shows that mitosis continues as the bacterium was guided through the microchannel creating an interwoven structure in which all of the nanofibers project in the same direction. The results in the Figure 6 clearly show that the orientation of cellulose fibers and the architecture of the network can be predictably controlled using electric fields.

Control of the 3D morphology of scaffolds from the nanoscale to the macroscale is a critical aspect in developing customizable implants and scaffolds for tissue engineering, since the biological responses of cells in these scaffolds is strongly correlated to the nanoscale topography[23]. We have demonstrated that this can be attained by controlling the movement of *A. xylinum* at the nanoscale. In this process the motion of *A. xylinum* is controlled by electric fields in a microfluidic environment to create custom cellulose networks. We have demonstrated the first direct control over a bottom up biofabrication process in three dimensions. By carefully controlling the electrokinetic forces, we can navigate the bacteria under simultaneous extrusion of cellulose networks. The manipulation of electrokinetic forces acting upon a bacterial cell can produce complex cellulose patterns on the nanoscale not achievable in static culture.

This offers, in addition to control of the material architecture, benefits from an environmental point of view. Our biofabrication process uses sugar and nutrients which can come from directly agricultural waste streams. Further, we do not use chemicals, operate at room temperate, use weak electric fields and the materials produced are biodegradable.

### *Conclusion*

The ability to control the direction of fiber orientation could be readily expanded to weave structures of multiple fiber layers, with each layer grown in a prescribed direction, by simply changing the orientation of the applied electric field, and these structures could be tailored to have the desired mechanical properties for a variety of applications including tissue engineering, MEMS, textiles, and electronics. Future work will focus on evaluating the mechanical properties of scaffolds with aligned fibers and creating scaffolds with complex fiber orientations suitable for biomedical implants. We are convinced that *A. xylinum* is not the only example of a biological system which can be controlled and the combination of multiple biological agents could be used to produce distinctive materials, connected at the nanoscale, which are morphologically and mechanically different from existing materials.

### *Acknowledgments*

We acknowledge the Institute for Critical Technologies and Applied Science (ICTAS) at Virginia Tech for financial support and the NCFL at ICTAS for support with imaging. We thank Dr. Aase Bodin for introducing Mike Sano to the bacterial cellulose field as well as Jaclyn Brennan, Phillip Zellner, Hadi Shafiee for experimental assistance. We also acknowledge Nathan Petersen and Michelle Davalos for assistance with editing.

### *References*

1. Wu, L.Q. and G.F. Payne, *Biofabrication: using biological materials and biocatalysts to construct nanostructured assemblies.* Trends in Biotechnology, 2004. **22**(11): p. 593-599.
2. Zareie, M.H., et al., *Controlled assembly of conducting monomers for molecular electronics.* Nano Letters, 2003. **3**(2): p. 139-142.
3. Zhou, H., et al., *Bacteria-based controlled assembly of metal chalcogenide hollow nanostructures with enhanced light-harvesting and photocatalytic properties.* Nanotechnology, 2009. **20**(8).
4. Lee, S.W., et al., *Ordering of quantum dots using genetically engineered viruses.* Science, 2002. **296**(5569): p. 892-895.
5. Brown, R.M., J.H.M. Willison, and C.L. Richardson, *Cellulose Biosynthesis In Acetobacter-Xylinum - Visualization Of Site Of Synthesis And Direct Measurement Of Invivo Process.* . Proceedings of the National Academy of Sciences of the United States of America, 1976. **73**(12): p. 4565-4569.

6. Brown, R.M., C. Haigler, and K. Cooper, *EXPERIMENTAL INDUCTION OF ALTERED NON-MICROFIBRILLAR CELLULOSE.* Science, 1982. **218**(4577): p. 1141-1142.
7. Backdahl, H., et al., *Mechanical properties of bacterial cellulose and interactions with smooth muscle cells.* Biomaterials, 2006. **27**(9): p. 2141-2149.
8. Watanabe, K., et al., *Structural features and properties of bacterial cellulose produced in agitated culture.* Cellulose, 1998. **5**(3): p. 187-200.
9. Borzani, W. and S.J. Desouza, *MECHANISM OF THE FILM THICKNESS INCREASING DURING THE BACTERIAL PRODUCTION OF CELLULOSE ON NON-AGITATED LIQUID-MEDIA.* Biotechnology Letters, 1995. **17**(11): p. 1271-1272.
10. Helenius, G., et al., *In vivo biocompatibility of bacterial cellulose.* Journal of Biomedical Materials Research Part A, 2006. **76A**(2): p. 431-438.
11. Klemm, D., et al., *Bacterial synthesized cellulose - artificial blood vessels for microsurgery.* Progress in Polymer Science, 2001. **26**(9): p. 1561-1603.
12. Svensson, A., et al., *Bacterial cellulose as a potential scaffold for tissue engineering of cartilage.* Biomaterials, 2005. **26**(4): p. 419-431.
13. Bodin, A., et al., *Bacterial cellulose as a potential meniscus implant.* Journal of Tissue Engineering and Regenerative Medicine, 2007. **1**(5): p. 406-408.
14. Bodin, A., et al., *Influence of cultivation conditions on mechanical and morphological properties of bacterial cellulose tubes.* Biotechnology and Bioengineering, 2007. **97**(2): p. 425-434.
15. Putra, A., et al., *Production of bacterial cellulose with well oriented fibril on PDMS substrate.* Polymer Journal, 2008. **40**(2): p. 137-142.
16. Putra, A., et al., *Tubular bacterial cellulose gel with oriented fibrils on the curved surface.* Polymer, 2008. **49**(7): p. 1885-1891.
17. Uraki, Y., et al., *Honeycomb-like architecture produced by living bacteria, Gluconacetobacter xylinus.* Carbohydrate Polymers, 2007. **69**(1): p. 1-6.
18. Kondo, T., et al., *Biodirected epitaxial nanodeposition of polymers on oriented macromolecular templates.* Proceedings of the National Academy of Sciences of the United States of America, 2002. **99**(22): p. 14008-14013.
19. Barz, D.P.J. and P. Ehrhard, *Model and verification of electrokinetic flow and transport in a micro-electrophoresis device.* Lab on a Chip, 2005. **5**(9): p. 949-958.
20. Tandon, V., et al., *zeta potential and electroosmotic mobility in microfluidic devices fabricated from hydrophobic polymers: 1. The origins of charge.* Electrophoresis, 2008. **29**(5): p. 1092-1101.
21. Kang, Y.J. and D.Q. Li, *Electrokinetic motion of particles and cells in microchannels.* Microfluidics and Nanofluidics, 2009. **6**(4): p. 431-460.
22. Son, H.J., et al., *Increased production of bacterial cellulose by Acetobacter sp V6 in synthetic media under shaking culture conditions.* Bioresource Technology, 2003. **86**(3): p. 215-219.
23. Stevens, M.M. and J.H. George, *Exploring and engineering the cell surface interface.* Science, 2005. **310**(5751): p. 1135-1138.

# Electrokinetic Manipulations of Microparticles in Curved Microchannels

**Junjie Zhu**, Mechanical Engineering, Clemson University, Clemson, SC

Tzuen-Rong Jeremy Tzeng, Biological Sciences, Clemson University, Clemson, SC

Xiangchun Xuan, Mechanical Engineering, Clemson University, Clemson, SC

We present electric field-mediated transport, focusing, and separation of microparticles in various curved microchannels. Due to the curvature-induced electric field non-uniformity, particles experience negative dielectrophoresis when they migrate electrokinetically through a curved microchannel. This cross-stream particle motion is utilized to either focus particles into a tight stream near the outer wall, or separate particles by size to dissimilar streams in spiral microchannels.

See more of this Session: Advances in Electrokinetics and Electrophoresis - Particles and Biomolecules
See more of this Group/Topical: Topical 3: 2009 Annual Meeting of the American Electrophoresis Society (AES)

# A NOVEL APPROACH TO DIELECTROPHORESIS USING CARBON ELECTRODES

*Rodrigo Martinez-Duarte and Marc. J. Madou*
*University of California, Irvine, Irvine, CA, USA*

## Introduction

Dielectrophoresis (DEP) enables the selective manipulation of a targeted particle, or population of particles, using the interaction of a non uniform electric field with the induced effective dipole moment of the particle(s) (1). DEP is a promising technique for the discrimination of particle populations and is amenable to miniaturization (2). It is advantageous over other particle separation techniques because discrimination between different particles is based solely on physical properties, *i. e.*, membrane morphology and internal compartmentalization in the case of cells. These physical properties determine the particle's dielectric properties and give it a characteristic dielectric phenotype. DEP works best when the dielectric phenotype of the targeted particle(s) significantly differs from those of other particles present in the sample(3). However, when operating in a flow system for continuous separation, populations with subtle differences can be separated by DEP but only at very low flow rates(4). In any case, the use of DEP eliminates the need for functionalized magnetic beads or fluorophores as needed by other separation techniques such as flow cytometry and magnetophoresis. The elimination of these often expensive labels reduces assay complexity, time and costs which can expand the availability of diagnostic tests, such as for HIV, and make therapies for cancer or degenerative diseases available to a broader number of patients. For example, DEP was demonstrated in the purification of blood and bone marrow from cancer cells (5-7) and in the enrichment of stem cell populations (8-11). For further reading, an excellent review on DEP fundamentals and the use of DEP in diagnostics is by Gascoyne and Vykoukal (12).

Most of the existent work in DEP relies on the polarization of metal microelectrodes to generate a non uniform electric field but this approach severely restricts the magnitude of the applied voltage between the electrodes as one must prevent sample electrolysis. Furthermore, the fabrication of micrometer-sized metal electrodes, especially when dealing with 3D electrodes rather than planar (2D) electrodes quickly turns relatively complicated and expensive. For example 3D metal electrodes require the use of metal electroplating that often restricts high yields and may result in more expensive devices. Dielectrophoresis can also be implemented using a technique known as insulator-based dielectrophoresis, or iDEP, where metal electrodes, ranging in size from millimeters to a couple of centimeters (extruded wire rods or machined plates for example), are positioned at both ends of an array of insulator structures (fabricated in glass or polymer) containing the sample. A high voltage is applied between the metal electrodes to generate a uniform electric field that is rendered non uniform at the surface of the insulator structures. Since the majority of the sample is only in contact with insulating materials the use of iDEP minimizes the possibility of electrolyzing the sample. However, iDEP requires very high voltages (electric field magnitude is inversely proportional to the gap between electrodes and in iDEP the separation between metal electrodes can be in the order of centimeters) to create a suitable electric field for dielectrophoresis.

Carbon-electrode dielectrophoresis, or carbonDEP, offers some key advantages over the above DEP techniques: 1) carbon electrodes feature a resistivity value (~1 X $10^{-4}$ $\Omega \cdot$ m (13)) that makes them conductive enough for the creation of a suitable electric field for dielectrophoresis using voltages in the range of tens of volts (in contrast to hundreds of volts needed in iDEP); 2) carbon has a much wider electrochemical stability window than gold or platinum and this affords higher polarizing voltages (and a stronger dielectrophoresis force induced on the targeted particle since DEP force is proportional to polarizing voltage square) in a given solution without electrolyzing it (14). Using the Carbon MEMS (C-MEMS) technique carbon electrodes, including 3D structures, are fabricated in a high yield process that is relatively simple and inexpensive (14, 15). The use of carbon electrodes yield other advantages such as excellent biocompatibility(16-18), chemically very inert in almost all solvents/electrolytes (19, 20) and good mechanical properties (21, 22).

Here we present the fabrication and filtering applications of selected carbon-electrode dielectrophoresis devices featuring volumetric (3D) carbon electrodes. Modeling and simulation work on carbonDEP is presented elsewhere (13, 23-25)

## Fabrication

3D Carbon-electrode DEP (carbonDEP) chips feature volumetric (3D) glass-like carbon electrodes that are obtained by the pyrolysis, or thermal degradation, of organic polymer structures using the C-MEMS technique (14). In this case, photopatterned SU-8 (an UV, epoxy-based negative photoresist) structures are used as the carbon precursor. The use of photolithography to pattern the carbon precursor allows for an exquisite dimension and shape control in the fabrication process of a wide variety of geometries. The fabrication process is summarized in Fig. 1. A two-step photolithography process is first carried on to fabricate the precursor SU-8 structures (Fig. 1a). These structures are then pyrolyzed at 900 °C under nitrogen atmosphere to yield an array of 3D carbon electrodes and their connecting electrical leads (the precursor polymer structures shrink isometrically during carbonization)(Fig. 1b). A thin SU-8 layer is patterned around the 3D carbon electrodes and on top of the connecting electrical leads to protect the leads from contact with the sample and other reagents such as buffers and cleaning agents (Fig. 1c). The micro channel required for the DEP chip is fabricated using polycarbonate and double-sided pressure-sensitive adhesive and is manually aligned to the array of 3D carbon electrodes. The fluid network is sealed by applying pressure on the SU-8/double-sided pressure-sensitive interface with a laminator. Fluidic connections to a syringe pump are implemented using commercial parts (Nanoport™ and PTFE tubing from Upchurch Scientific). Electrical contact from the carbon electrodes and their connecting electrical leads to a function generator is made with simple wiring and using a silver/solder interface on the connection pad (coating the carbon pad with silver paint and soldering a wire to the silver). The experimental device and a cross-section of the active DEP area featuring 3D carbon electrodes are illustrated in Fig. 2a and b respectively.

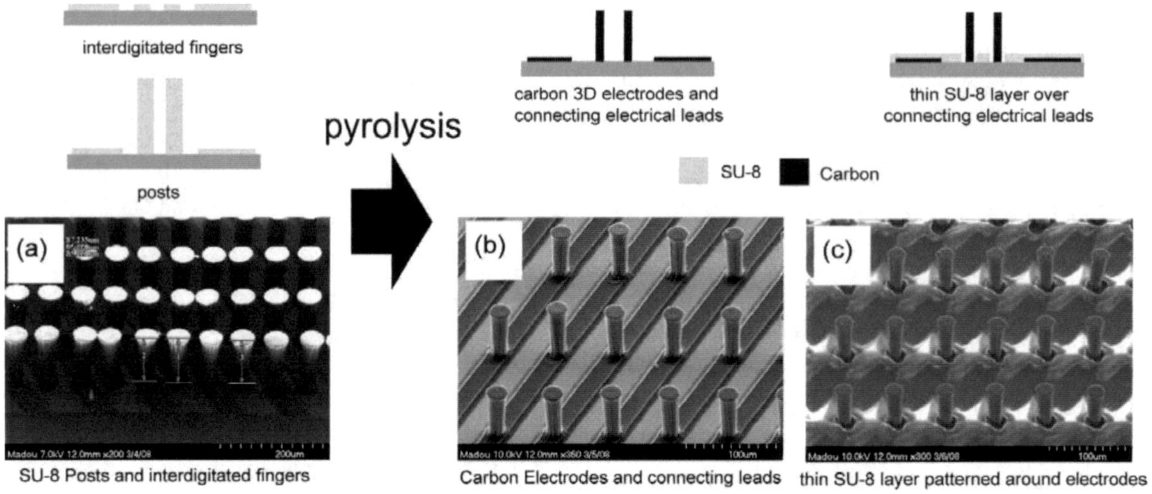


**Fig. 1** Carbon-electrode DEP device fabrication

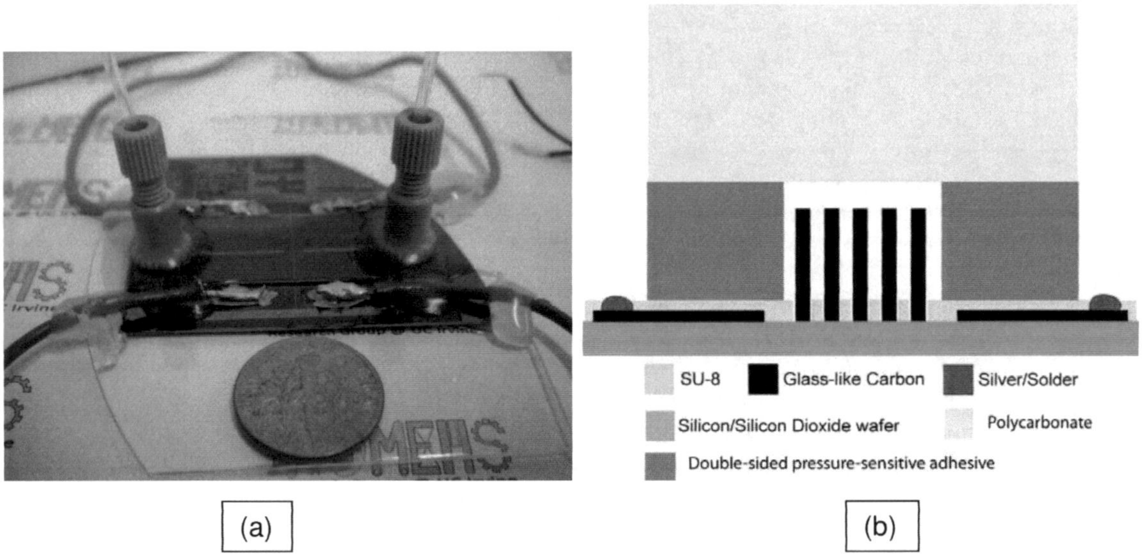


**Fig. 2** a) Experimental device and its b) cross-section

## Filter Applications

The use of 3D electrodes that cover most of the height of a flow channel improves filter throughput by reducing the mean distance of any particle to the closest electrode surface and reduces or eliminates the number of re-flow cycles required to improve filtering when using more traditional planar (2D) electrodes as illustrated in Fig. 3.

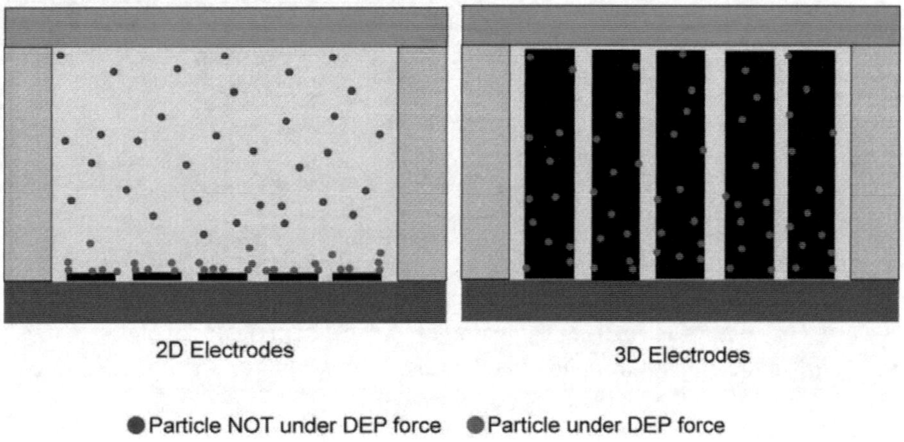


**Fig. 3** 2D vs. 3D approach

The improvement on filter efficiency and throughput when using 3D electrodes instead of 2D electrodes was quantified by Martinez-Duarte et al. (26) when filtering viable yeast cells from non viable ones. The experimental sample was a mixture of viable and non viable yeast cells suspended in a conductive media and it was flowed over the polarized electrodes at different flow rates (in the 2-20 μl/minute range). The retention efficiency as a percentage, calculated as the ratio of the quantity of the targeted particles (viable cells) retained by the filter, to the quantity entering the filter, for different flow rates is shown in Fig 4a. The use of 3D electrodes yields a better filter; with higher retention efficiency than that achieved by 2D electrodes at all flow rates. The performance of the 3D electrode array diminishes as flow rate increases but the decrease in performance is not nearly as sharp as that of 2D electrodes. The filtering of viable yeast cells from non viable yeast ones when flowing the experimental sample over the electrodes at a flow rate of 10 μl per minute in a channel with a rectangular cross-section (600 by 100 μm) is illustrated in Fig. 4b. Viable yeast cells are trapped in those areas with the highest electric field magnitude as predicted by electromagnetic simulation (24, 25).

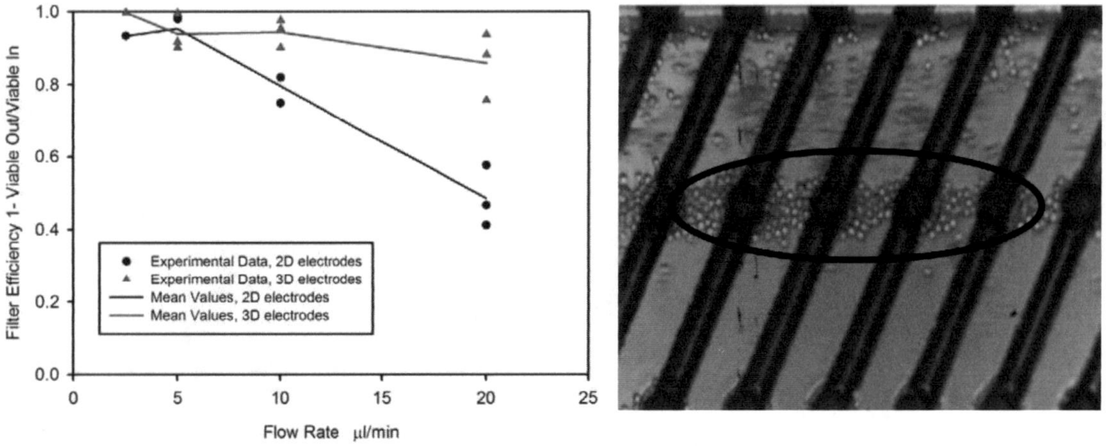


(a) (b)

**Fig. 4** a) Filter efficiency comparison between 2D (blue) and 3D (red) as flow rate increases. b) Filtering of viable yeast cells (in the ellipse) from non viable yeast cells when sample flows at a rate of 10 μl/min over the polarized electrodes

A two-stage filter was also demonstrated by Martinez-Duarte et al. (27, 27). Two electrically independent 3D electrode arrays are excited by two different sinusoidal signals (each optimized on their magnitude and frequency to selectively trap targeted particles in a particular array). This approach was demonstrated by filtering viable yeast cells in one array and non viable yeast cells in another electrode array from an experimental sample containing latex particles, viable and non viable yeast cells. The selective filtering of yeast cells on different electrode arrays is illustrated in Fig. 5.

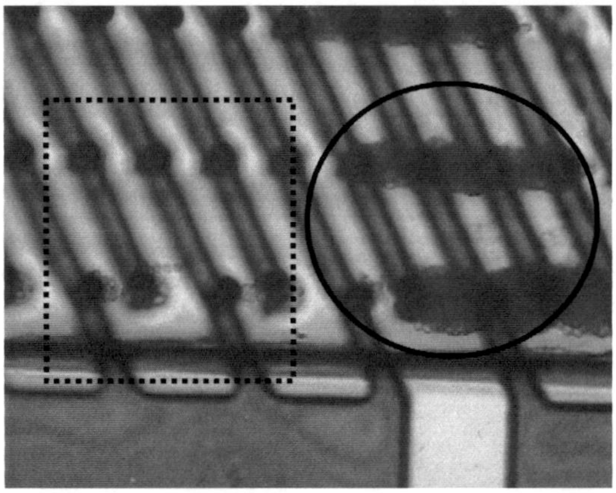

**Fig. 5** Two-stage filter featuring two independent carbonDEP arrays. Each array is excited by different sinusoidal signals. Viable yeast cells are trapped in the left array (dashed square) while non viable yeast cells are trapped in the right array (solid circle)

## Conclusion

Carbon-electrode dielectrophoresis, or carbonDEP, is an alternative to the use of metal electrodes or insulator structures (iDEP) in dielectrophoresis applications. Compared to metal electrodes, glass-like carbon has a wider electrochemical stability window than gold and platinum which reduces the possibility of electrolyzing the sample. The fabrication of carbon 3D structures using the C-MEMS technique is significantly simpler than the fabrication of 3D metal structures with techniques such as electroplating. Compared to iDEP, the use of 3D carbon electrodes yields throughputs similar to those achieved with iDEP but carbon electrodes are conductive enough to only require tens of volts to create a suitable electric field for DEP (instead of the hundreds of volts required with iDEP).

## References


1. Pohl,H. A. (1978), "Dielectrophoresis," Cambridge University Press: Cambridge.
2. Castellanos,A., Ramos,A., González,A., Green,N. G. and Morgan,H. (2003), "Electrohydrodynamics and dielectrophoresis in microsystems: scaling laws," *J.Phys.D,* 36, pp. 2584-2597.
3. Gascoyne,Peter R. C., Wang,Xiao-Bo, Huang,Ying and Becker,Frederick F. (1997), "Dielectrophoretic Separation of Cancer Cells from Blood," *IEEE T Ind Appl,* 33, pp. 670-678.
4. Menachery,A., Chappell,S., Errington,R. J., et al. (2008), "Dielectrophoretic characterization and separation of metastatic variants of small cell lung cancer cells," *in Proceedings of NSTI Nanotech 2008:* Boston, MA, June 1-5, pp. 386-389.
5. Gascoyne,P. R. C., Wang,X. B., Huang,Y. and Becker,F. F. (1997), "Dielectrophoretic separation of cancer cells from blood," *IEEE Trans.Ind.Appl.,* 33, pp. 670-678.
6. Becker,F. F., Wang,X. -., Huang,Y., Pethig,R., Vykoukal,J. and Gascoyne,P. R. C. (1995), "Separation of human breast cancer cells from blood by differential dielectric affinity," *Proc.Natl.Acad.Sci.U.S.A.,* 92, pp. 860-864.
7. Yang,J., Huang,Y., Wang,X. B., Becker,F. F. and Gascoyne,P. R. C. (1999), "Cell separation on microfabricated electrodes using dielectrophoretic/gravitational field flow fractionation," *Anal.Chem.,* 71, pp. 911-918.
8. Flanagan,L. A., Lu,J., Wang,L., et al. (2008), "Unique Dielectric Properties Distinguishing Stem Cells and Their Differentiated Progeny," *Stem Cells,* 26, pp. 656-665.
9. Stephens,M., Talary,M. S., Pethig,R., Burnett,A. K. and Mills,K. I. (1996), "The dielectrophoresis enrichment of CD34+ cells from peripheral blood stem cells harvests," *Bone Marrow Transplantation,* 18, pp. 777-782.
10. Huang,Y., Yang,J., Wang,X-B, Becker,F. F. and Gascoyne,P. R. C. (1999), "The Removal of Human Breast Cancer Cells from Hematopoietic CD34+ Stem Cells by Dielectrophoretic Field-Flow-Fractionation," *Journal of Hematotherapy & Stem Cell Research,* 8, pp. 481-490.
11. Talary,M. S., Mills,K. I., Hoy,T., Burnett,A. K. and Pethig,R. (1995), "Dielectrophoretic separation and enrichment of CD34+ cell subpopulation from bone marrow and peripheral blood stem cells," *Medical and Biological Engineering and Computing,* 33, pp. 235-237.
12. Gascoyne,P. R. C. and Vykoukal,J. V. (2004), "Dielectrophoresis-based sample handling in general-purpose programmable diagnostic instruments," *Proc IEEE,* 92, pp. 22-42.

13. Park,B. Y., Taherabadi,L., Wang,C., Zoval,J. and Madou,M. J. (2005), "Electrical Properties and Shrinkage of Carbonized Photoresist Films and the Implications for Carbon Microelectromechanical Systems Devices in Conductive Media," *J Electrochem Soc,* 152, pp. J136-J143.
14. Wang,Chunlei, Jia,Guangyao, Taherabadi,Lili H. and Madou,Marc J. (2005), "A Novel Method for the Fabrication of High-Aspect Ratio C-MEMS Structures," *J of MEMS,* 14, pp. 348-358.
15. Wang,Chunlei and Madou,Marc. (2005), "From MEMS to NEMS with carbon," *Biosens and Bioelectron,* 20, pp. 2181-2187.
16. Benson,J. (1969), "Carbon offers advantages as implant material in human body," *NASA Technical Brief,*
17. Benson,J. (1971), "Elemental Carbon as a Biomaterial," *J Biomed Mat Res,* 5, pp. 41-47.
18. Turon Teixidor,G., Gorkin III,R. A., Tripathi,P. P., et al. (2008), "Carbon microelectromechanical systems as a substratum for cell growth," *Biomed Mater,* 3, pp. 034116.
19. Yamada,S. and Sato,H. (1962), "Some Physical Properties of Glassy Carbon," *Nature,* 193, pp. 261-262.
20. Cowlard,F. C. and Lewis,J. C. (1967), "Vitreous Carbon - A New Form of Carbon," *J Mat Sci,* 2, pp. 507-512.
21. Kotlensky,W. V. and Martens,H. E. (1965), "Tensile Properties of Glassy Carbon to 2,900 C," *Nature,* 206, pp. 1246-1247.
22. Jenkins,G. M. and Kawamura,K. (1971), "Structure of Glassy Carbon," *Nature,* 231, pp. 175-176.
23. Park,B. Y. and Madou,M. J. (2005), "3-D electrode designs for flow-through dielectrophoretic systems," *Electrophoresis,* 26, pp. 3745-3757.
24. Martinez-Duarte,R., Cito,S., Collado-Arredondo,E., Martinez,S. O. and Madou,M. (2008), "Fluido-dynamic and Electromagnetic Characterization of 3D Carbon Dielectrophoresis with Finite Element Analysis," *in Proceedings of NSTI Nanotech 2008:* Boston, MA, pp. 265-268.
25. Martinez-Duarte,R., Cito,S., Collado-Arredondo,E., Martinez,S. O. and Madou,M. J. (2008), "Fluido-dynamic and Electromagnetic Characterization of 3D Carbon Dielectrophoresis with Finite Element Analysis," *Sens and Transd J,* 3, pp. 25-36.
26. Martinez-Duarte,R., Rouabah,H. A., Green,N. G., Madou,M. and Morgan,H. (2007), "Higher Efficiency and Throughput in Particle Separation with 3D Dielectrophoresis with C-MEMS," *in Proceedings of uTAS 2007:* Paris, France, Oct. 7-11, pp. 826-828.
27. Martinez-Duarte,R., Andrade-Roman,J., Martinez,S. O. and Madou,Marc J. (2008), "A High Throughput Multi-stage, Multi-frequency Filter and Separation Device based on Carbon Dielectrophoresis," *in Proceedings of NSTI Nanotech 2008:* Boston, MA, June 1-5, pp. 316-319.

# Microfluidic Continuous Particle Separation BASED ON Electrical Properties Via AC-Dielectrophoresis

**Barbaros Cetin**, Mechanical Engineering Department, Vanderbilt University, Nashville, TN

Dongqing Li, Mechanical and Mechatronics Engineering, University of Waterloo, Waterloo, ON, Canada

In this study, we are presenting a novel, simple microfluidic based lab-on-a-chip device for the continuous separation of particles according to their electrical properties by using AC-Dielectrophoresis. The device is composed of a straight rectangular channel connecting two inlet reservoirs with the two exit reservoirs. The main flow is induced by the pressure difference between the inlet and the exit reservoirs. The dielectrophoretic separation is achieved by a hybrid design using a PDMS (poly-dimethylsiloxane) hurdle and a pair of embedded metal electrodes to generate localized non-uniform AC electric field. The hurdle geometry is optimized to get the best performance for the separation. Particle trajectories are simulated by using Lagrangian tracking method and the experimental verification of the simulations are performed by using latex particles.

See more of this Session: Advances in Electrokinetics and Electrophoresis - Particles and Biomolecules
See more of this Group/Topical: Topical 3: 2009 Annual Meeting of the American Electrophoresis Society (AES)

# Continuous Colloidal Concentration and Patterning with Optically Induced AC Electrokinetics

Stuart J. Williams, Birck Nanotechnology Center, Purdue University, West Lafayette, IN

Aloke Kumar, Birck Nanotechnology Center, Purdue University, West Lafayette, IN

Nicolas G. Green, School of Electronics and Computer Science, University of Southampton, Southampton, United Kingdom

Steven T. Wereley, Birck Nanotechnology Center, Purdue University, West Lafayette, IN

We demonstrate an optically induced AC electrokinetic technique that rapidly and continuously accumulates colloids (49 nm to 3.0 microns) on both ITO and gold electrode surfaces resulting in a crystalline-like monolayer aggregation. This is done without photosensitive materials. Electrothermal hydrodynamics produce a microfluidic vortex that carries particles in suspension towards its center where they are trapped by low-frequency AC electrokinetic forces. We characterize the rate of particle aggregation as a function of the applied AC voltage, AC frequency, and illumination intensity. Further, we have selectively captured and separated 2.0 micron from 1.0 micron polystyrene particles as well as 1.0 micron silica from 1.0 micron polystyrene particles. This dynamic optically-induced technique rapidly and selectively concentrates, patterns, and translates colloidal aggregates and can be used for a variety of lab-on-a-chip applications.

See more of this Session: Advances in Electrokinetics and Electrophoresis - Particles and Biomolecules
See more of this Group/Topical: Topical 3: 2009 Annual Meeting of the American Electrophoresis Society (AES)

# Pressure-Generation at the Intersection of Two Microchannels with Different Depths Via Electrokinetic Means

**Naoki Yanagisawa**, Chemistry, University of Wyoming, Laramie, WY

Debashis Dutta, Chemistry, University of Wyoming, Laramie, WY

While fluid and analyte samples are often driven through microfluidic channels using electroosmotic flow, pressure-driven transport is preferred in several applications due to its ability to move a broader range of materials and insensitivity to channel interfacial properties. Key among them is the implementation of liquid chromatography where this mode of actuation significantly improves the reproducibility and the controllability of the separations by decoupling mobile phase transport from analyte interaction with the stationary phase. Generation of pressure-driven flow on microchip devices with a high degree of control however, offers a significant challenge. Although conventional syringe pumps may be connected to microchips to realize steady hydrodynamic flows, the dynamic control over the transport rates in these systems is poor due to the large dead volumes involved.

Here, we report the design of a microchip based hydraulic pump that can generate pressure-gradients within a microchannel network via electrokinetic means allowing very precise dynamic control over the flow rates. The pump consists of 3 glass channel segments in a tee geometry; one of which is shallower (about 1-2mm deep) than the other two (about 5-15mm deep). By applying appropriate electric fields across this design, a mismatch in electroosmotic transport rate is introduced at the junction of the shallower and the deeper microchannels which occurs due to a differential in the effective fluid conductivity in these two segments. For example, it has been observed that the effective fluid conductivity in glass microchannels increases with a reduction in the channel depth due to an increasing contribution from surface conductance [1]. The pressure-driven flow generated in the system as a result of this mismatch is then guided to a separation section in this device using the third segment in the tee geometry. With our current design, pressure-driven flow velocities up to 5mm/s have been generated in the separation column for an applied voltage of 2kV. In addition, experiments have been also performed to understand the effect of the buffer ion concentration and the relative depth of the shallower channel segment in this design on the pressure-generation capability of our device. Finally, the functionality of our device has been demonstrated by performing an open channel liquid chromatography of fluorescent dyes under pressure-driven flow conditions in a 5mm separation channel.

References:

[1] Stein, D., Kruithof M, and Dekker C, "Surface charged governed ion transport in nanofluidic channels", Phys. Rev. Lett. 2004, 93, Article No: 035901

See more of this Session: Advances in Electrokinetics and Electrophoresis - Particles and Biomolecules
See more of this Group/Topical: Topical 3: 2009 Annual Meeting of the American Electrophoresis Society (AES)

## Dielectrophoretic separation of Nano-particle conjugated bacterial cells within micro-scale architecture

*Swarnasri Mandal[1], Rajeev Kumar Singh[1], Deepak Singh[2], R. Gurunath[2], Shantanu Bhattacharya [1*]*

*(1)Department of Mechanical Engineering (2) Department of Chemistry, Indian Institute of Technology, Kanpur, India. *Corresponding Author's Email: bhattacs@iitk.ac.in*

### ABSTRACT:

The detection and counting of bacterial cells can provide a tremendous impetus to rapid diagnostics and clinical applications. The traditional culture, growth and colony counting techniques, although well-established and reliable, are time-consuming (may need a few hours to days) and thus, cannot provide a fast and accurate means of quantitation which may be vital in some emergency situations [1]. In addition, these traditional methods need extra effort and investment for maintenance of sterility for accurate and reproducible results. The utilization of chip-based techniques has obvious advantages in terms of enhanced speed of detection, automation, and point-of-care use. [2–7] For this purpose a concentration step is normally added prior to the detection process. One of the most widely used concentration techniques in fluidic microchips is dielectrophoresis (DEP). This refers to the interaction of non-uniform electric field with polarization charges induced in any electrically neutral particle which results in differential forces and selective trapping over the electrodes causing the non homogenous electric field.[8] Dielectrophoresis has emerged as a very useful tool for selective trapping and sample pre-concentration. For alternating fields the selectivity towards a certain particle of a specific dielectric constant depends on the operating frequency of such an alternating field. We are developing a novel cell sorting method using nanoparticle conjugates of bacterial cells by actively capturing conjugated cells over a set of interdigitated electrodes and then also measuring the capture efficiency. We start with a mixture of E.Coli (BL21) and Psuedomonas Aerogenosa cells. Goat anti mouse IgG conjugated gold nano-particles (10nm) (Ted Pella Inc., Redding, CA, USA) are used to form conjugates with E.Coli (BL21) cells using monoclonals of Anti E. Coli antibodies.[9] Finally, the gold conjugated cells in the mixture sample are flown through a hybrid micro-fluidic platform with an upper die of PDMS, containing a micro-channel and a lower die of silicon with imprinted interdigitated electrodes. We intend to separate the nano-particle conjugated E. Coli (BL21) cells from the mixture. The capture yields will be verified by fluorescence measurements and the selectivity will be evaluated by running a multiplex PCR over the retentate solution

### INTRODUCTION:

Bacteria existed long before humans evolved, and bacterial diseases probably co-evolved with each species which involuntarily hosts them for the longest time we were not aware of the cause of infectious diseases. To detect pathogens causing food borne diseases at an early stage was a major challenge since long time. However techniques based on culture enrichment & plate count are available to detect pathogens but it takes 3-7 days for result. Thus rapid detection techniques become vital to food safety. Recent development in this direction is achieved by Liju Yang et al. They developed a microfluidic device in which pathogen were concentrated by using DEP. The selective capture of Listeria monocytogenes from sample was achieved by biotinylated BSA–streptavidin–biotinylated monoclonal antibody sandwich structure [9]. Ning bao developed the microfluidic chip capable of detecting autofluorescence from bacterial cell lysate for quantitative determination of the cell number. The autofluorescence is present in all cells and does not require labeling. This method is very useful when number of cells is well below the detection limit. The laser-induced autofluorescence can thus quantitatively detect the cells if they are as low as 210 in the sample [12].

However detection, selection and identification have always tried to invoke the inquisitive mind. With the advent of DEP on microchips the issue of preconcentration of lean samples and their counting by various means including optical [10], electrical [13] etc. was achieved. Their identification was also subsequently performed extensively by use of antibody based recognition [9] and molecular methods like PCR [11]. However, there is still a need of selection from group of bacteria of a particular type. This need is very prominent in current day diagnostics as it opens up a new methodology for rapid detection of blood borne pathogens which in any case are in traces and can cause huge immunogenic responses. We use DEP to our advantage further inspired by the work reported by Soh et. al. and try to make nanoconjugates of bacteria (E. Coli BL21) and trying to exploit their changed electrical properties.

**EXPERIMENT AND METHODS:**

*Fabrication of bonded device made of Silicon and PDMS*

The device was realized in two parts viz. upper die and lower die and bonded irreversibly by plasma bonding (Plasma Cleaner PDC- 32G, Harrick Plasma).

*Realization of upper die*

Microchannels are made in upper die by replication and molding method. The material selected for making upper die was PDMS (Polydimethylsiloxane). Being optically transparent it offers the great advantage of image acquisition during sample flow. The SU-8 structures corresponding to microchannels were printed on silicon wafer using photolithography. The photoresist was selectively exposed using black transparency mask. The unexposed areas were etched away with developer. The minimum feature size depends upon the film thickness which can be controlled by spin speed. These microstructures were used as mold for replication purpose. The PDMS prepolymer and its curing agent were mixed in a ratio of 10:1 and poured on the mold. This mixed polymer (liquid at room temperature) was degassed using desiccator & then cured in an oven at $85^0$C for 45 mins. This matrix can be easily separated by peel off technique from the SU-8 structures. The microchannels were imprinted on PDMS due to corresponding microstructures on SU-8. Figure 1 shows the cross-sectional dimensions of the various parts of the micro-channel imaged with a Nikon Eclipse 80i fluorescence microscope. Two holes were drilled at the both ends of PDMS channel over the broader areas for inlet and outlet of the fluid. Micro PEEK tubing ¼ in OD, 200 microns ID (M/s Upchurch Scientific) is thereby used by inserting the ends in the drilled holes on the inlet and outlet ports and thus form a continuous flow system.

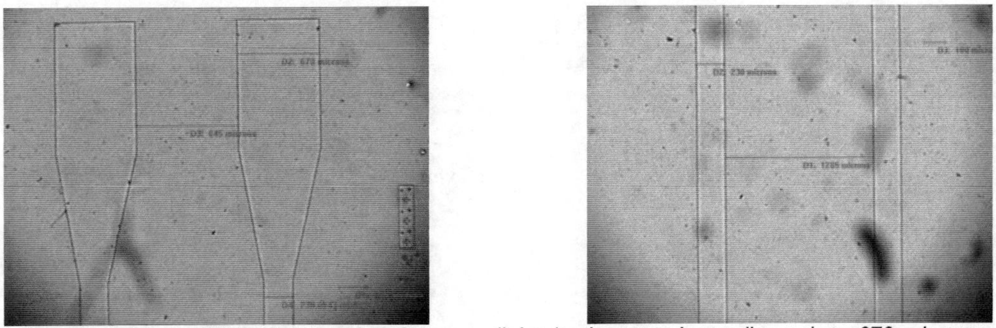

Figure 1: Images taken at 4X for inlet and outlet ports (inlet /outlet ports have dimensions 670 microns and the main portion of the channel have dimensions 230 microns)

*Realization of the lower die:*

A silicon wafer is well cleaned using standard protocol and spin coated with a positive tone resist (S-1813, M/s Shipley). This photoresist is selectively exposed to UV using a hard mask. To create the mask, a design package, AutoCAD is used. The mask was made in chrome and glass. The exposed photoresist were etched away by developer thus creating vias and trenches. These vias and trenches were exposed to metal vapors. A layer of 20 nm of titanium & 200 nm platinum were sputter coated in these trenches and vias. Then unexposed part of resist was removed. In this way, the interdigitated electrodes were realized. Figure 2 shows the microscopic images of such electrodes. Each device was made on a piece of silicon around 1"×1" .

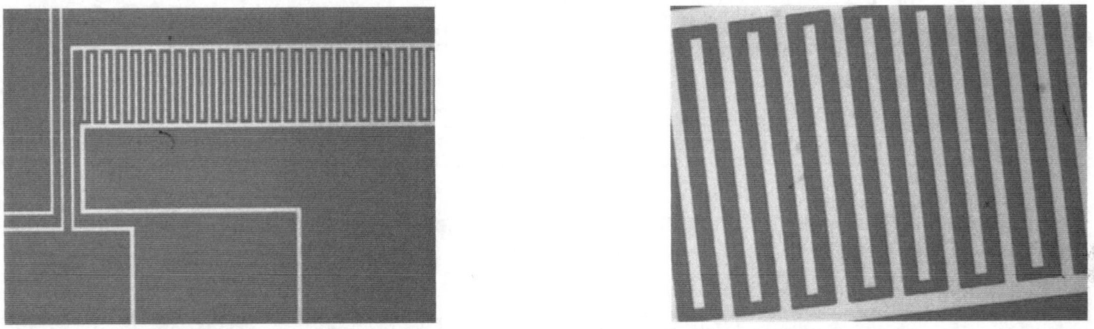

Figure 2: (a) Image of the electrodes taken at 10X mag (b) Image of the electrodes at 100X mag (the electrodes are separated from each other by an inter electrode spacing of 20 microns and their individual width is around 17 microns)

*Assembly of both the dies*

The microchannels made in PDMS are open from the lower end & this need to be assembled to interdigitated electrodes in order to flow. In order to bond two dies, oxygen plasma bonding is used. While bonding, the surface properties changes with interaction to oxygen plasma. Due to change in surface energy, irreversible bonding between surfaces takes place. The biggest advantage of this process is that no intermediate glue is required between two surfaces. The assembly is schematically shown in figure 3 (a) and (b).

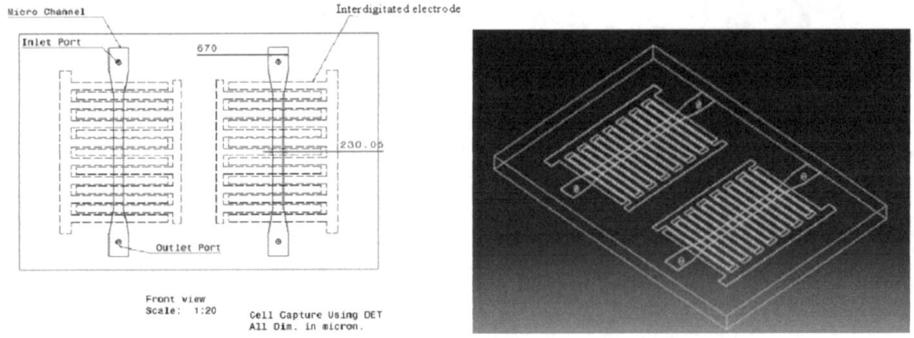


Figure 3: (a) Schematic representation of the Plan view of the device (b) Isometric view of the final assembly

*Sample Preparation*

The samples are prepared by taking a fresh culture. Bacterial cell were put to fresh culture media for about 12 hrs prior to experiment (OD of about 2.5 is considered suitable to have considerable number of cells). Firstly the cells are centrifuged and washed to remove the media and then dyed using acridine orange; 3µg/ml for 15 mins. The culture is once again centrifuged and washed with phosphate buffer multiple times in order to minimize the background fluorescence.

A part of the culture realized in the above step is mixed with anti E.coli antibody solution (0.1mg/ml) (Abcam ab25823). This mixture is left to incubate for 15 mins following which goat anti-mouse IgG (heavy and light chain) conjugated with 10 nm gold particles (Ted Pella, Inc., Redding, CA, USA) is added (1 : 50 dilution)and kept for 10–15 min at room temperature. This strategy is developed to nanoconjugate the bacterial cells with the gold nanoparticles. The SEM images of both the normal and conjugated E.coli BL21 cells are taken using Carl Zeiss evo 50.

*Capturing bacterial cells using DEP*

The E.coli BL21 cells are captured using A/C fields applied by means of two probe tips and an A/C signal generator (M/s Agilent technologies). A 10V peak to peak with a sinusoidal nature is given to the interdigitated electrodes while flowing the bacterial cells at a rate of 0.1ml/hr (Harvard® 11 Plus dual syringe pump) through the microchannels mounted just over these. We observe that a large number of bacterial cells are captured while flowing through the applied electric field as we apply a frequency of 10K Hz.

*Fluorescence Imaging*

The DEP captured images for the fluorescently labeled bacteria are acquired using Nikon ECLIPSE 80i fluorescence microscope (Japan) using a TRITC (tetramethylrhodamine isothiocyanate) filter (Excitation wavelength 540-565nm),

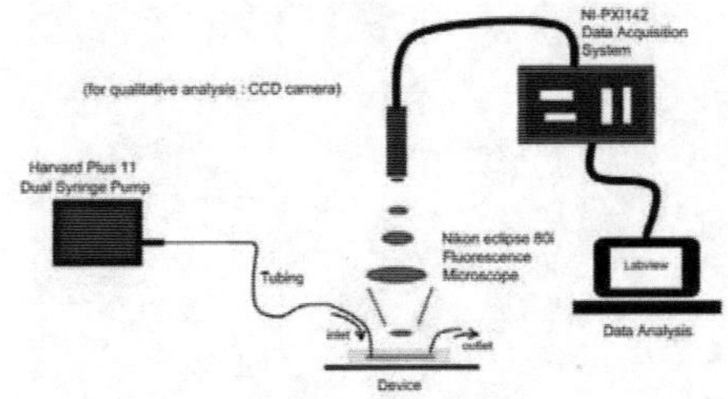


Figure4: Schematic representation of experimental setup

**RESULT AND DISCUSSION:**

*Characterization of Dielectrophoretic force*

According to following equation of Dielectrophoresis:

$$\left\langle \vec{F}(t) \right\rangle = 2\pi\epsilon_0\epsilon_m R^3 Re\left[f_{CM}\right] \vec{\nabla}\left(E_{RMS}\right)^2$$

The Dielectrophoretic force depends on the electric field gradient which implies geometry and configuration of electrodes producing the field. Electrode edges have the maximum electric field due to the coiling of field lines around sharp corners. So the bacterial cells are expected to collect at the edges in positive DEP at low frequency A/C signals. As the frequency slowly increases conductivity increases and permittivity decreases. We can also calculate the DEP force by varying the voltage applied at the electrode.

$$\varepsilon^* = \varepsilon - j\frac{\sigma}{\omega}$$

*Bacterial flow with and without DEP*

Figure 5 (a)-(b) represents static images of the microchannels with E.coli BL21 over electrodes in the presence and absence of electric fields.

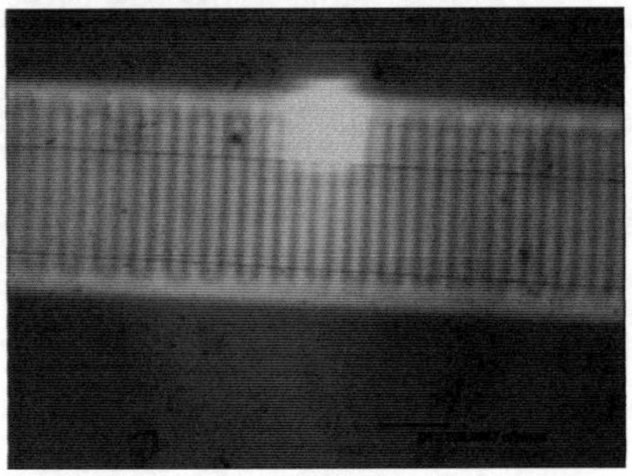

(a)

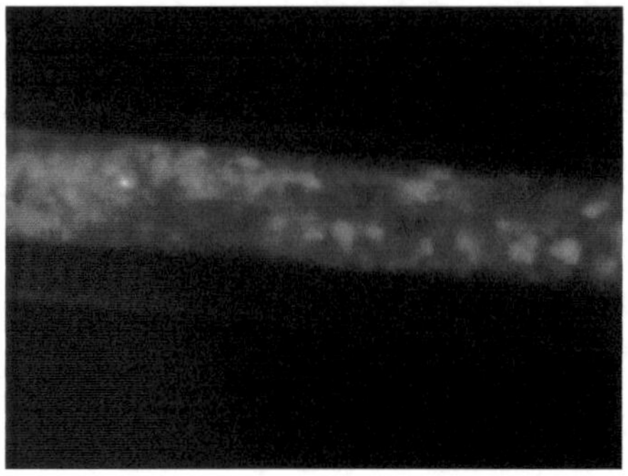

(b)

Figure 5: (a) Without DEP (b) Capture of E.coli with DEP

It has been observed that the extent of capture is higher near the inlet port i.e. at initial stages of the flow. The portion of the electrodes interacting apriori with the flow gets more no. of bacteria than the portion later in the flow path as the concentration of bacteria reduces due to the initial capture. In order to estimate the capture efficiency of DEP we perform plate culture of both the initial solution and post DEP solution obtained from the outlet. The number of bacterial colonies decreases by orders of magnitude in the post DEP sample compared to the pre DEP sample(Figure 6 (a & b)).

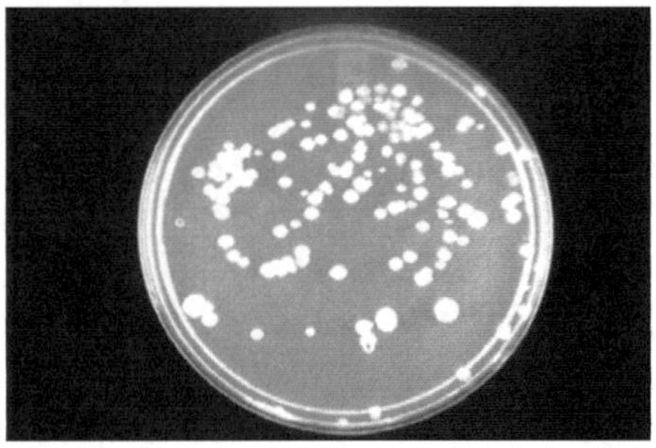

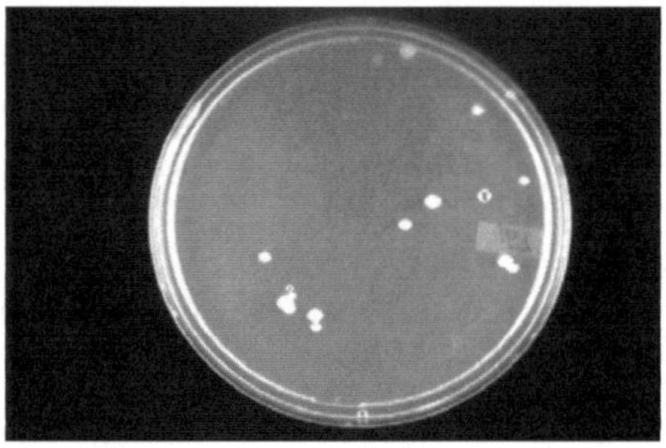

(a) (b)

Figure 6: Gel Doc images (a) Pre DEP sample (b) Post DEP sample

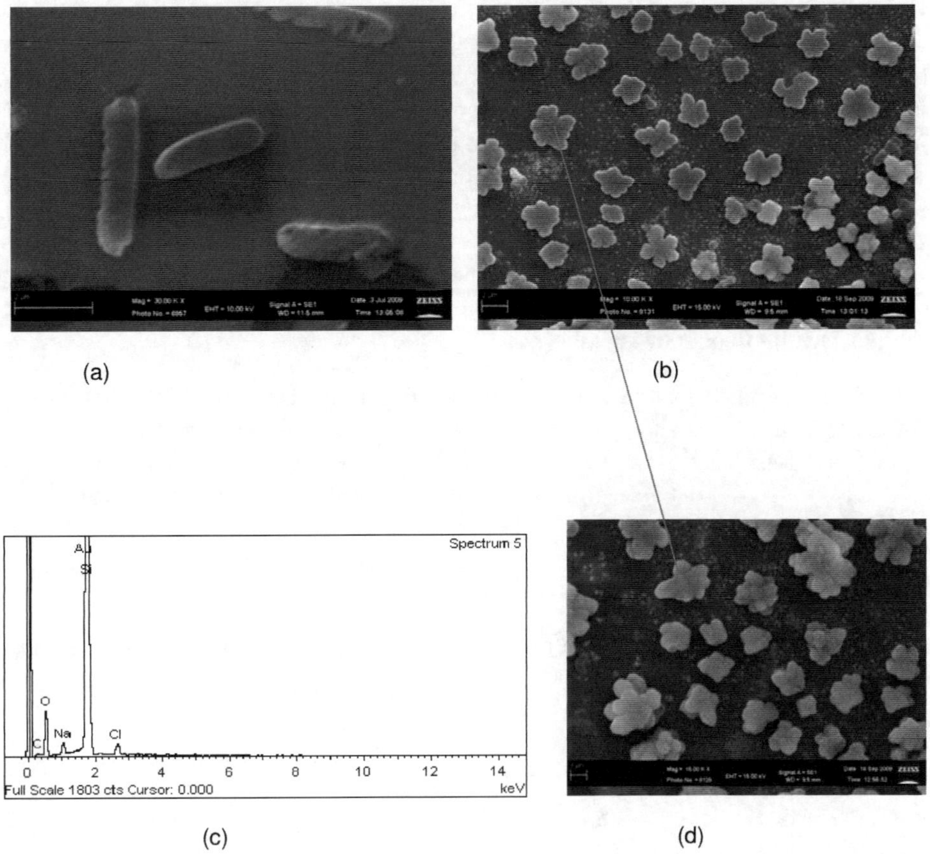


Figure 7: (a) SEM image of unconjugated E Coli BL21 cells (b and d) SEM images of conjugated cells at 10,000 X and 15,000 X (c) EDX spectra of a 30 micron X 30 micron area showing the prominence of gold.

From the SEM images of the conjugated bacteria we see that they have formed clumps which can be attributed to the presence of monoclonal. However, what is also very interesting to see is the formation of borders of high luminosity around all the conjugates. They probably signify the presence of gold resulting in lustrous glow. We have also performed an Energy Dispersive X-ray analysis (EDX) on a 30X30 micron $^2$ area and it contains peaks for O, C, Na, K, Cl along with Au as indicated in figure 7.

We are currently trying to perform DEP on the conjugated cells which due to the change in shapes and also possibly the dielectric constant, has a tendency to get captured at much higher frequencies (approximately 1 MHz) as opposed to the capture of normal unconjugated cells at 10KHz. We are currently completing and developing an understanding of this separation technique. Futher we would like to perform these studies from a standpoint of separation of only the conjugated specimens from a lot of mixed bacterial culture. We would like to validate our data by doing PCR and verifying the accuracy of separation and selectivity.

## CONCLUSION:

In this study, we have demonstrated a microfluidic device which captures bacterial cells by DEP at a particular frequency. We also try to conjugate the bacterial cells with antibody and goat anti-mouse IgG attached to gold nanoparticles. On account of this conjugation the bacterial cells have been found to have altered shapes and sizes and also show the conjugation to gold particles as seen on an SEM by lustrous borders. The confirmation of the presence of gold is further visible by the appearance of a gold peak in the EDX data. Thus we believe that such an effect will lead to an altered capture frequency on performing DEP, hence enabling differential capture of conjugated and nonconjugated E.coli cells. At this time the frequency of capture is found to be more and we are trying to develop an understanding of why the frequency is highered.

We further intend to extend such a method to capture cells of different species at various ranges of frequencies and estimating the efficiency using fluorescence measurement followed by PCR analysis of the retentate solution .

## REFERENCES:


[1]Identification of Microbial Activity, www.corrosionsource.com /technicallibrary/ corrdoctors/Modules /Microbial/Identification.htm

[2] S. S. Iqbal, M. W. Mayo, J. G. Bruno, B. V. Bronk, C. A. Batt and J. P. Chambers, *Biosens. Bioelectron.*, 2000, **15**, 549–578.

[3] M. G. Giminska, *Microbial Cell Factories*, 2006, **5**, 1–8.

[4] M. Ramsey, *Nat. Biotechnol.*, 1998, **16**, 40–49.

[5] C. F. Edman, P. Mehta, R. Press, C. A. Spargo, G. T.Walker and M. Nererberg, *J. Invest. Med.*, 2000, **48**, 93–101.

[6] T. M. H. Lee, M. C. Carles and I. M. Hsig, *Lab Chip*, 2003, **3**, 100–105.

[7] E. T. Lagally, P. C. Simpson and R. A. Mathies, *Sens. Actuators, B*, 2000, **63**, 138–146.

[8] Joel Voldman, Rebecca A. Braff, Mehmet Toner, Martha L. Gray, Martin A., Schmidt. *A.,* Van den Berg A., *Micro Total Analysis Systems 2000*, 431-434.

[9] L. Yang, P. Banada, M.R. Chatni, K.S. Lim, A.K. Bhunia, M. Ladisch, R. Bashir, *Lab chip*, 2006, **6**, 896-905.

[10] Anil Ghubade, Swarnasri Mandal, Rahul Chaudhury, Rajeev Kumar Singh and Shantanu Bhattacharya, Biomedical microdevices , 10.1007/s10544-009-9316-6.

[11] Shantanu Bhattacharya, Shuaib Salamat, Dallas Morisette, Padmapriya Banada, Demir Akin, Yi-Shao Liu, Arun K. Bhunia, Michael Ladisch and Rashid Bashir, Lab Chip, 2008, 8, 1130 – 1136.

[12] Ning Bao, Balarnurugan Jagadeesan, Arun K. Bhunia, Yuan Yao, Chang Lu, Journal of Chromatography A, 2008, 1181 153–158.

[13] Rafael Gómez-Sjöberg, Dallas T. Morisette, Rashid Bashir, Journal of Microelectromechanical Systems, 2005, 14(4), 829-838.

# Free-Solution Conjugate Electrophoresis Sequencing of Over 250 Bases of ssDNA

**Jennifer A. Coyne**, Chemical Engineering, Stanford University, Stanford, CA

Jennifer S. Lin, Bioengineering, Stanford University, Stanford, CA

Annelise E. Barron, Bioengineering, Stanford University, Stanford, CA

The list of genomes that have been or will be studied by sequencing continues to grow, with applications ranging from agriculture to molecular medicine and understanding cancer. Next-generation sequencing methods are enabling a majority of these ultra-high throughput projects by lowering the cost and time associated with genomic sequencing studies. However, a need remains for low cost, targeted, high-accuracy sequencing of limited medical samples for which rapid electrophoresis-based methods are ideal. Significant research has been done to translate electrophoretic separations onto microfluidic devices, resulting most recently in successful sequencing of 600 bases in 6.5 minutes by chip electrophoresis. The advantages of microfluidic separations for these targeted medical sequencing applications (faster read length, smaller sample volume, potential integration into lab-on-a-chip devices), however, are dampened by the difficulties associated with loading the viscous polymer solutions necessary to achieve size-based separations of sequencing fragments. **Free-Solution Conjugate Electrophoresis (FSCE)** enables size-based separations with no entangled polymer network through the specific conjugation of exactly identical mobility modifiers to each DNA molecule, which breaks the linear scaling of the charge and friction of DNA with its length. In FSCE, these completely monodisperse, mostly charge-neutral perturbing entities ("drag-tags") are chemically conjugated to the 5'-end of ssDNA oligomers or sequencing primers. An ideal drag-tag, in addition to being completely monodisperse, must be uniquely attached to DNA and large in size in order to provide long and accurate reads; genetically engineered, highly repetitive "protein polymers" have been developed for this use. Recent advances in FSCE drag-tags and separations have improved the sequencing length to over 250 bases of DNA separated in free-solution (using a 267-amino acid protein polymer), which is approximately a 40% increase in read-length over previously published FSCE sequencing results and is comparable to the read-length of next-generation sequencing technologies. A drag-tag with 516-amino acids has recently been produced that is almost completely monodisperse, and sequencing with this drag-tag is underway. Two examples of medical sequencing where FSCE is applicable are the donor-recipient matching process for organ transplants, where 40 HLA regions < 450 bases must be sequenced, and the detection and identification of nucleotide mutations in individual exons can affect the prognosis, diagnosis and treatment of cancer.

See more of this Session: DNA Analysis in Microfluidic & Nanofluidic Devices
See more of this Group/Topical: Topical 3: 2009 Annual Meeting of the American Electrophoresis

Abstract: Free-Solution Conjugate Electrophoresis Sequencing of Over 250 Bases of ssDNA (2009 Annual Meeting)

Society (AES)

# Nanoporous Polymer Membrane for DNA Preconcentration Enables Sensitive Chip Electrophoresis

Robert J. Meagher, Sandia National Laboratories, Livermore, CA

Anup K. Singh, Biosystems Research and Development, Sandia National Laboratories, Livermore, CA

Conventional chip electrophoresis using a double-T injection design enables high peak efficiency due to the narrow injection plug, but detection sensitivity is often limited because this format offers relatively little sample stacking, and much of the sample is wasted simply filling the injection area. Numerous other approaches have been demonstrated for on-line preconcentration of DNA samples, including isotachophoresis, micropatterned electrodes, and sequence-specific "capture matrices". Large concentration factors are possible, but each technique presents certain drawbacks in terms of simplicity of device fabrication or operation.

We report on the use of thin (~30 micron) photopatterned polymer membranes for on-line preconcentration of single- or double-stranded DNA samples prior to electrophoretic analysis. Shaped UV laser light is used to quickly (~10 seconds) polymerize a narrow, highly crosslinked polyacrylamide plug. By applying an electric field across this membrane, DNA from a dilute sample can be concentrated into a narrow zone (<100 micron wide) at the outside edge of the membrane. The field at the membrane can then be reversed, allowing the narrow plug to be cleanly injected into a separation channel filled with a sieving polymer for analysis. Concentration factors >100 are possible, increasing the sensitivity of analysis for dilute samples. We have fabricated both neutral membranes (purely size-based exclusion) as well as anionic membranes (size and charge exclusion), and characterized the rate of preconcentration as well as the efficiency of injection from both types of membrane, for several sizes of DNA, ranging from a 20 base ssDNA oligonucleotide to >14 kbp dsDNA. We have also investigated the effects of concentration polarization on device performance for the charged membrane. Results for sensitivity as well as separation efficiency are compared to those for the Agilent Bioanalyzer, a widespread and successful platform for chip electrophoresis.

Advantages of the membrane preconcentration approach include the simplicity of device fabrication and operation, and the generic (non-sequence specific) nature of DNA capture, which is useful for complex or poorly characterized samples where a specific capture sequence is not present. The membrane preconcentration approach is well suited to simple single-level etch glass chips, with no need for patterned electrodes, integrated heaters, valves, or other elements requiring more complex chip fabrication.

See more of this Session: DNA Analysis in Microfluidic & Nanofluidic Devices

See more of this Group/Topical: Topical 3: 2009 Annual Meeting of the American Electrophoresis Society (AES)

# Microchip-Based Investigation of the Interplay Between the Nanoporous Gel Morphology and the Onset of Entropic Trapping in DNA Gel Electrophoresis

**Nan Shi**, Artie McFerrin Department of Chemical Engineering, Texas A&M University, College Station, TX

Victor M. Ugaz, Artie McFerrin Department of Chemical Engineering, Texas A&M University, College Station, TX

The design of miniaturized DNA electrophoresis systems imposes demanding requirements that are challenging our knowledge about how the nanoscale pore morphology of the sieving matrix influences the physical mechanism of DNA migration through the gel. But a detailed understanding of this interplay is limited by a general lack of experiments probing fundamental parameters like mobility and diffusion coefficients that are key descriptors of electrophoretic migration. We have recently developed an innovative way to address this need through the use of a microfluidic platform that allows continuous whole channel scanning of DNA separation progress so that the size dependence of these parameters can be quickly and accurately measured. When combined with material characterization methods we have developed that enable both the mean pore size and pore size distribution of the gel to be quantified, we are able to establish a direct link between sieving matrix morphology and DNA migration in photopolymerized crosslinked polyacrylamide gels cast under different UV intensities and concentrations.

We have applied this approach is applied to examine electrophoresis of double-stranded DNA with lengths ranging from 0.1 to 1 kb in these gels, and have discovered that varying the rate of photopolymerization appears to induce a corresponding change in the physical mechanism of DNA migration between reptation and entropic trapping. Furthermore, we have identified a range of conditions where a favorable interplay exists between these migration mechanisms that results in improved separation performance. These counterintuitive results have led us to reconsider the applicability of some scaling relationships and theories widely used to describe electrophoretic migration of DNA, and we present our progress toward developing modifications that can enable them to account for the influence of the gel pore size and pore size distribution. This will allow us to better predict separation performance in different sieving gels and operating conditions, and ultimately provide new insights that can help enable rational selection of optimal matrix materials and polymerization conditions that deliver enhanced separation performance.

See more of this Session: DNA Analysis in Microfluidic & Nanofluidic Devices

See more of this Group/Topical: Topical 3: 2009 Annual Meeting of the American Electrophoresis Society (AES)

# DNA Collisions with Polarizable Posts

**Daniel W. Trahan**, Chemical Engineering, Massachusetts Institute of Technology, Cambridge, MA

Patrick S. Doyle, Chemical Engineering, Massachusetts Institute of Technology, Cambridge, MA

We have studied electrophoretic collisions of DNA with an ideally polarizable post. By ideally polarizable, we mean that the dielectric constant of the post $\varepsilon_p$ is much greater than that of the surrounding fluid $\varepsilon_f$ (i.e., $\varepsilon_p >> \varepsilon_f$). This situation is very different from previously studied post collision problems that either neglected the field disturbances due to the obstacle [1-2] ($\varepsilon_p = \varepsilon_f$) or only considered the insulating case [3] ($\varepsilon_p << \varepsilon_f$). For an insulating post, the electric field lines all move around the post which tends to hinder significant polymer collision events except in the case of a molecule approaching the post along the centerline. Polarizable posts, on the other hand, attract and focus the field lines which should lead to strong polymer/post interactions.

Molecular collisions with obstacles have attracted a great deal of interest recently due to their potential use in macromolecular separations [4]. In the case of DNA, collisions often lead to hooking events where a molecule drapes itself around a post with an arm stretched around each side of the post. The molecule then unhooks by one of the arms growing at the expense of the other. The time required for this unhooking process is length dependent which provides a separation mechanism that has been exploited using large post arrays. DNA separation has been demonstrated in both experiments [4] and simulations [5, 6].

More broadly speaking, molecular collisions are important because they provide a simple and convenient way to manipulate individual molecules in both applications and studies relating to fundamental science. For example, obstacle collisions have been exploited in DNA mapping devices to "precondition" molecules for stretching in an elongational field to overcome the effects of molecular individualism [7]. And, more fundamentally, obstacles have been used to deform DNA to study the relaxation process of polymers in confined environments [8].

The electric field around an ideally polarizable post has no tangential component at the post surface, so the field lines intersect the post at right angles. On the upstream side of the post, the field lines point into the post while on the downstream side they point away. As a molecule approaches, it is driven into the post, which it cannot penetrate. Due to the finite size of the molecule (i.e., its center of mass is located slightly off the post surface), the field does provide a weak tangential component to move it around to the downstream side. Therefore, the molecule should only slowly move around the post's surface until it crosses to the downstream side where the field lines begin to emanate from the post. The downstream field can then pull the rest of the molecule around the post, at which point it can freely convect away.

When the molecule is driven into the post on the upstream side, it is compressed causing the molecule's radial dimension $R_p$ to decrease. This length determines the molecule's exposure to the tangential field that exists off the surface of the post, and thus, it also governs the speed at which the molecule moves around the post. Based on this insight, we have developed scalings for the field-driven compression of a molecule against a solid surface, which lead to scalings for the time a molecule requires to clear a post.

We used blob theory to derive our scalings for $R_p$. Based on this framework, we were able to balance the entropic and enthalpic penalties due to compression against the resulting reduction of the molecule's electrical potential. Finally, we have also performed Brownian dynamics simulations in order to validate our models and more fully understand the behavior of the molecular collisions.

References:


[1] G. I. Nixon and G. W. Slater, Phys. Rev. E, 50, 5033 (1994).

[2] P. M. Saville and E. M. Sevick, Macromolecules, 32, 892 (1999).

[3] G. C. Randall and P. S. Doyle, Phys. Rev. Lett., 93, 058102 (2004).

[4] P. S. Doyle, J. Bibette, A. Bancaud, and J. L. Viovy, Science, 295, 227 (2002).

[5] P. D. Patel and E. S. G. Shaqfeh, J. Chem. Phys., 118, 2941 (2003).

[6] A. Mohan and P. S. Doyle, Phys. Rev. E, 76, 040903 (2007).

[7] A. Balducci and P. S. Doyle, Macromolecules, 41, 5485 (2008).

[8] W. D. Volkmuth, T. Duke, M. C. Wu, R. H. Austin, and A. Szabo, Phys. Rev. Lett., 72, 2117 (1994).


See more of this Session: DNA Analysis in Microfluidic & Nanofluidic Devices
See more of this Group/Topical: Topical 3: 2009 Annual Meeting of the American Electrophoresis Society (AES)

# Free-Energy Landscaping: Tailoring DNA Dynamics Across a Fluidic Nanotopography

**Derek M. Stein**, Physics, Brown University, Providence, RI

Jackson Travis Del Bonis-O'Donnell, Physics, Brown University, Providence, RI

Walter Reisner, Physics, Brown University, Providence, RI

The pressure-driven transport of DNA was studied in slit-like nanochannels with an embedded nanotopography consisting of linear arrays of nanopits. We imaged individual DNA molecules moving single-file down the nanopit array, undergoing sequential pit-to-pit hops using fluorescence video microscopy. Distinct transport dynamics were observed depending on whether a molecule could occupy a single pit, or was forced to subtend multiple pits. We interpret these results in terms of a scaling theory of the free energy of polymer chains in a linear array of pits. Molecules contained within a single pit are predicted to face an entropic free energy barrier, and to hop between pits stochastically by thermally activated transport. Molecules that subtend multiple pits, on the other hand, can transfer DNA contour from upstream to downstream pits in response to an applied fluid flow, which lowers the energy barrier. When the trailing pit completely empties, or when the leading pit reaches its capacity, the energy barrier is predicted to vanish, and the low-pressure, thermally activated transport regime gives way to a high-pressure, deterministic transport regime. These results contribute to our understanding of polymers in nanoconfined environments, and can guide the design of nanoscale lab-on-a-chip applications for DNA analysis.

See more of this Session: DNA Analysis in Microfluidic & Nanofluidic Devices
See more of this Group/Topical: Topical 3: 2009 Annual Meeting of the American Electrophoresis Society (AES)

# Electrokinetically Driven DNA Dynamics in Nanochannel/Microwell Array

**Orin L. Hemminger**, Chemical and Biomolecular Engineering, Ohio State University, Columbus, OH

Pouyan E. Boukany, Nanoscale Science and Engineering Center for Affordable Nanoengineering of Polymeric Biomedical Devices, The Ohio State University, Columbus,, OH

Xin Hu, NSEC center for Affordable Nanoengineering of Polymer Biomedical Devices (CANPBD), The Ohio State University, Columbus, OH

Nan-rong Chiou, Department of Chemical and Biomolecular Engineering, The Ohio State University, Columbus, OH

Ly James Lee, The Ohio State University, Columbus, OH

Recently, we developed a simple and low-cost method to produce polymeric nanochannel/microwell arrays, which can be easily connected to microscale inlets/outlets to form a nanofluidic device. Such device has potential for gene delivery and DNA separation. The nanochannel/microwell array was fabricated in ethylene glycol dimethylacrylate (EGDMA). Hydrophobicity/Hydrophilicity and zeta potential of the channel wall can be adjusted by acid treatment. The periodic length of nanochannels ranged from 1 to 10 um with diameter of 10-200 nm. The microwells with different size and shape can be used to pre-stretch large DNA molecules and to adjust the electric field strength in nanochannels. Using spin-disk dynamic confocal microscopy, we investigate the transport behavior of various sizes of DNAs under different periodic nanoscale confinements by direct flow visualization, upon applications of pre-specified electric field strengths. The experimental observations are compared with coarse-grain Brownian dynamic simulation.

See more of this Session: DNA Analysis in Microfluidic & Nanofluidic Devices
See more of this Group/Topical: Topical 3: 2009 Annual Meeting of the American Electrophoresis Society (AES)

## Modulating DNA Adsorption On Silica Beads in An Electrically Actuated Microfluidic Device

*Tao Geng, Ning Bao, Oren Z. Gall, Chang Lu*[1]
Purdue University, West Lafayette, IN, USA

### Introduction

DNA adsorption/desorption on solid surfaces serves as the underlying mechanism for the vast majority of DNA preconcentration and separation techniques [1-7] that are available for genetic analysis of biological samples [8-10]. Silica surfaces are the most common ones that are applied for DNA extraction and concentration [1-3, 5, 11]. DNA interaction with silica surface has been intensively studied to illustrate mechanistic details of these adsorption/desorption processes [11-18]. In the majority of reported protocols, DNA extraction from biological samples was performed by inducing DNA binding onto silica surface in a high ionic strength buffer in the presence of high concentration salts (e.g. 6M Guanidium HCl) and later eluting DNA out in a low ionic strength buffer [1-3, 5, 11, 16, 17].

It is known that pH has a strong influence on the adsorption of DNA on silica surface [12, 15, 18]. In general, high pH decreases DNA adsorption on silica due to electrostatic repulsion (both the surface and DNA are negatively charged). As pH gets lower, the electrostatic repulsion disappears because the silica surface is protonated and the hydrophobic interaction and hydrogen bonding facilitate the adsorption of DNA molecules. In this study, we present a strategy for electrical actuation of DNA adsorption on silica beads on a microfluidic chip. In our design, we apply electricity through surface electrodes to electrochemically vary pH in a packed bed of silica beads. The pH change alters the amount of DNA molecules adsorbed on silica beads. In this way, we are able to switch the state of the packed bed from being absorptive to desorptive without changing the buffer. Our approach provides a novel method for regulating DNA adsorption by electricity in a low ionic strength buffer.

### Materials and Methods

#### Device design

The design of the electrically actuated microfluidic device is illustrated in Fig. 1. Gold electrodes were fabricated on the glass substrate to carry out electrolysis of water [19, 20]. The two gold electrodes (200 μm wide) were 1000 μm apart from each other. The device consists of an inlet channel (20 μm deep), a "weir" (3 μm deep) and a outlet channel (20 μm deep). The square pillars were fabricated in the weir to prevent it from collapsing onto the glass substrate (shown in the inset image in Fig. 1). Due to the depth variation, microscale beads with the diameter larger than the weir depth could be trapped by the weir to form a packed bed, as shown in the inset image in Fig. 1. The packed bed was located at

[1] Author to whom correspondence should be addressed; email: changlu@purdue.edu

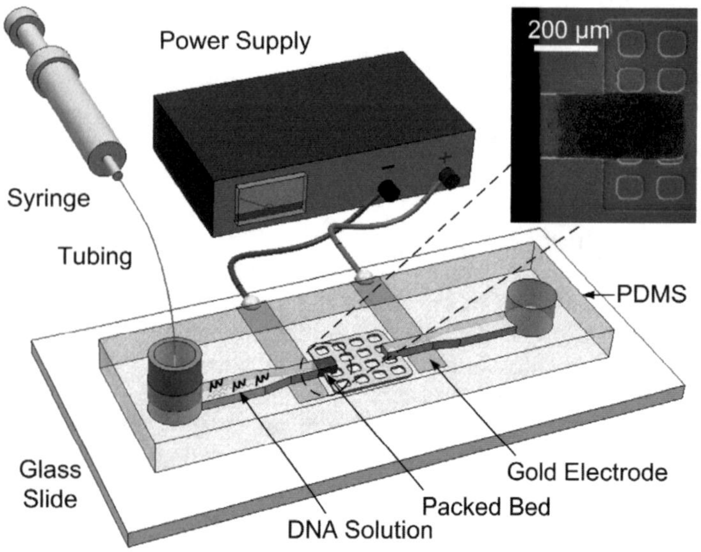


**Fig. 1.** The design of the electrically actuated DNA adsorption device. The inset image shows the weir, the packed bed of silica beads, the inlet channel, and the gold microelectrode (cathode). Note: The drawing is not to scale.

the downstream of the cathode. A syringe pump was used to sweep the hydroxide generated at the cathode to the packed bed of silica beads and increase pH there when the voltage between the two electrodes was turned on. The packed bed was desorptive when the voltage was on and adsorptive when the voltage was off. The applied voltage affects the rate of hydroxide generation. Generally, higher voltage introduces more electrolysis of water which changes the pH more rapidly. However, we found that a voltage higher than 23 V generated visible bubbles at the cathode. With the voltage of 22 V we applied, we observed no apparent bubbling and had steady electrochemical reaction.

### Device fabrication

The device was fabricated using standard soft lithography technique [21, 22]. Two photomasks were used to fabricated the weir and the rest of the device sequentially on the same silicon wafer to form the master. The weir was fabricated in a negative photoresist (SU-8 2002, MicroChem Corp., Newton, MA) and the other parts were done using a positive photoresist (AZ 9260, Clariant Corp., Somerville, NJ) with the two sections aligned properly during the photolithography. A PDMS chip was molded using the master. To fabricate microelectrodes, an adhesion layer of titanium (20 nm thick) followed by a layer of gold (150 nm) was deposited onto the pre-cleaned glass slides using an E-beam evaporator (Temescal FC-1800, Fairfield, CA). The microelectrodes were then fabricated by wet etching. The PDMS chip and the cleaned glass slide (with microelectrodes) were oxidized in a plasma cleaner (Harrick Plasma, Ithaca, NY), aligned, and bonded together. Finally, the whole device was baked at 80 ºC for 2 hrs to ensure strong bonding between PDMS and the glass slide.

### Buffer preparation

Buffers of pH 4.0, 5.0, 6.0 and 7.0 were prepared by mixing 1 M of acetic acid (HAc), 1 M of sodium acetate (NaAc) and 1 M of sodium chloride (NaCl), as shown in Table 1. The buffer of pH 10.0 was prepared by mixing 25 mL of 0.1 M $Na_2HPO_4$ solution and 0.87 mL of 0.1 M NaOH and then diluting to 100 mL.

**Table 1. Preparation of 25 mM acetate buffer of different pHs**

| pH | 1.0 M HAc (μL) | 1.0 M NaAc (μL) | 1.0 M NaCl (μL) | $H_2O$ (mL) |
|---|---|---|---|---|
| 4.0 | 2120 | 400 | 2120 | 95.4 |
| 5.0 | 900 | 1600 | 900 | 96.6 |
| 6.0 | 133 | 2400 | 133 | 97.4 |
| 7.0 | 14 | 2500 | 14 | 97.5 |

### DNA labeling

λ-DNA (48,502 base pairs, Sigma, St. Louis, MO) was stocked in 10 mM Tris-HCl buffer (pH 7.0). DNA samples were fluorescently labeled with YOYO-1 (Invitrogen, Carlsbad, CA) at a ratio of one dye molecule per five base pairs with a concentration of 1 nM in acetate and phosphate buffer solutions. After incubating the mixture at room temperature for at least 1 hr in the dark, these samples were further diluted to have a DNA concentration of 100 pM with proper buffer solutions immediately prior to the experiment.

### Device operation

The inlet of the device was first connected to a syringe pump (PHD infusion pump, Harvard Apparatus, Hollistion, MA) through plastic tubing. A suspension of silica microbeads (4.8 μm diameter, Bangs laboratories, Fishers, IN, or 10 μm diameter, Corpuscular, Cold Spring, NY) with a concentration of 0.05% was subsequently loaded into the channel at 3 μL/min. After the beads packed to a length of ~400 μm in the microfluidic channel, DNA sample flowed into the channel at different flow rates (0.3 μL/min, 1 μL/min, 2 μL/min and 3 μL/min) for the study of electrical actuation of DNA adsorption. To apply electricity, microelectrodes were connected to a direct current power supply with a constant voltage of 22 V. The packed bed was stable during the experimental runs. All experiments were conducted at 25 ± 1 ºC in the dark.

### Fluorescence imaging

For fluorescence intensity measurements, the microfluidic device was mounted on an inverted fluorescence microscope (IX-71, Olympus, Melville, NY) equipped with a 20× dry objective and a CCD camera (ORCA-285, Hamamatsu, Bridgewater, NJ). The epifluorescence excitation was provided by a 100 W mercury lamp. The excitation and emission from λ-DNA labeled with YOYO-1 were filtered by a fluorescence filter cube (Exciter HQ480/40, emitter HQ535/50, and beamsplitter Q505lp, Chroma Technology, Rockingham, VT). One image was taken every two min. To avoid photobleaching, the shutter was normally closed, and opened for less than two seconds during image capturing. Images were processed using ImageJ (NIH, http://rsb.info.nih.gov/ij) software to acquire the fluorescence intensity data.

## Results and Discusstion

Previous DNA adsorption studies were mostly conducted on fused silica substrates [12,18]. We first uses our deivce to verify the effect of the buffer pH on DNA adsorption on the surface of silica beads. 25 mM acetate buffers with different pHs (pH 4.0, 5.0, 6.0 and

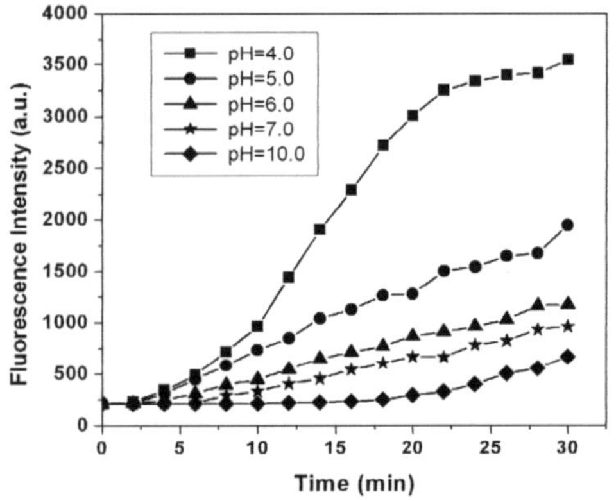


**Fig. 2** The adsorption of DNA molecules onto silica beads (4.8 μm diameter) in 25 mM acetate buffers (pH 4.0, 5.0, 6.0 and 7.0) and 25 mM phosphate buffer (pH 10.0) at a flow rate of 1 μL/min. No electricity was involved in this experiment.

7.0) and 25 mM phosphate buffer (pH 10.0) were used for the study. While λ-DNA molecules (100 pM) in buffers of different pHs was continuously pumped through a packed bed of silica beads with a flow rate of 1 μL/min, fluorescence intensity from YOYO-1 labeled DNA molecules absorbed on the bead surface was measured as a function of time. It needs to be noted that the intercalating dye YOYO-1 (positively charged) decreases the negative charge of DNA molecules as observed previously [23]. Nevertheless, we expect DNA molecules without labeling to have similar behaviors in our device. The results in Fig. 2 indicate that the surface concentration of adsorbed DNA molecules increased with time and reducing pH (also see Fig. 3). In particular, DNA molecules were much more dramatically adsorbed at pH 4.0 than at other pHs and the adsorption was largely suppressed at pH 10.0. These results are consistent with previous studies on silica substrates [12].

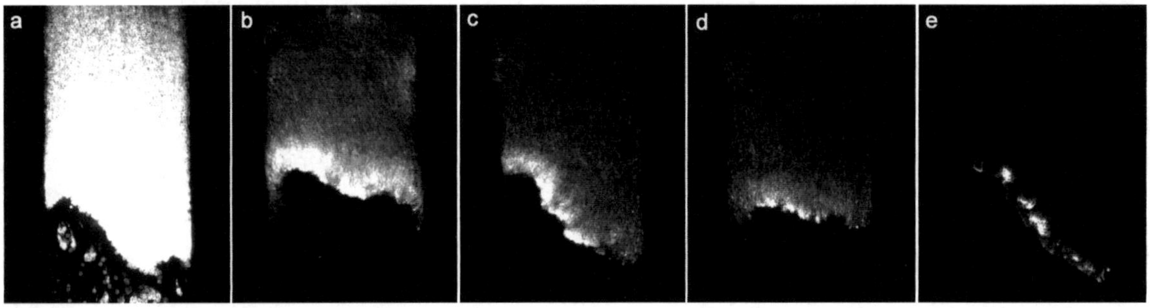


**Fig. 3** Fluorescent images of DNA molecules adsorbed on silica beads (4.8 μm diameter) in (a) pH = 4.0, (b) pH = 5.0, (c) pH = 6.0, (d) pH = 7.0, and (e) pH = 10.0 buffers at a flow rate of 1 μL/min. Images were captured after the DNA solution flowed for 20 mins in the channel with a exposure time of 100 ms.

We subsequently tested the electrical modulation of DNA adsorption on silica beads with three flow rates for the DNA solution with pH 7.0. In Fig. 4a, 100 pM DNA solution in 25 mM acetate buffer (pH 7.0) flowed into the device with a rate of 1 μL/min during the entire course of the experiment. With a pH of 7.0 in the buffer, DNA molecules were first adsorbed on the bead surface and the fluorescence intensity from the packed bed increased continuously. Then we switched on the constant voltage (22 V) between the electrodes and a drop in the fluorescence intensity was immediately observed. It took about 20 min for the fluorescence intensity to drop to its minimum and start to plateau. An increase in the fluorescence intensity was seen again shortly after removal of the voltage, due to the re-establishment of the original buffer pH with fresh DNA-containing buffer flowing in. The accumulation of DNA molecules on the beads considerably slowed down compared to the first time adsorption. The bare surface of silica is weakly acidic due to the

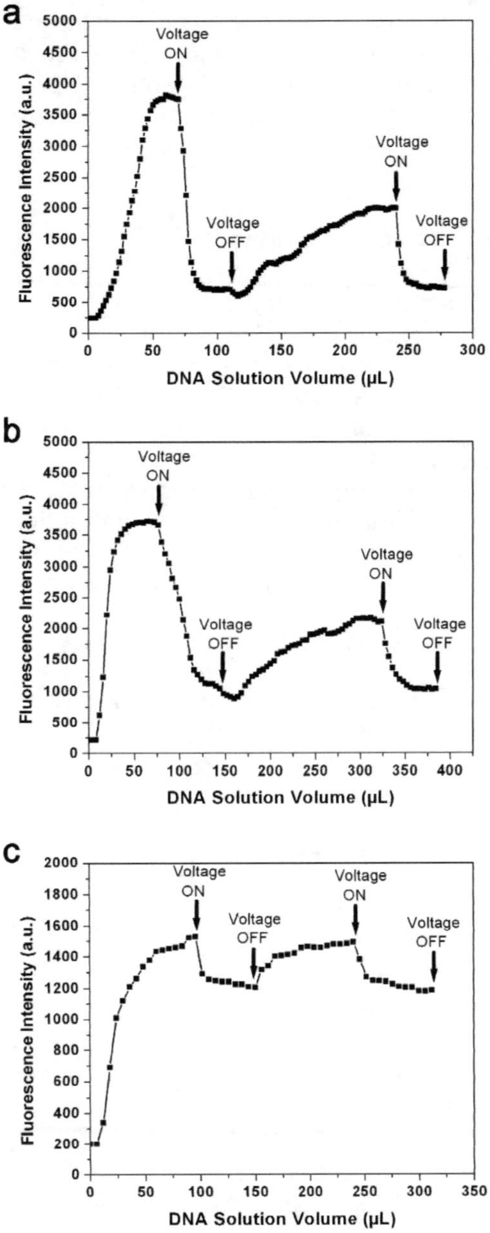


**Fig. 4** The modulation of DNA adsorption on silica beads (10 μm diameter) by electricity in 25 mM acetate buffer solution (pH 7.0) at a flow rate of (a) 1 μL/min, (b) 2 μL/min, and (c) 3 μL/min. The arrows show the onset and the removal of the constant voltage (22 V). Exposure time for each fluorescent image was 100 ms for (a, b) and 10 ms for (c).

silanol groups with a $pK_a$ ranging from 5.0 to 7.0. A higher effective $pK_a$ of 7.5 is reported for the surface silanol groups interacting with adsorbed DNA [17]. DNA molecules adsorbed on silica beads when the buffer pH was 7. When the electrochemical reaction started to generate hydroxide in the upstream at the cathode and the pH in the packed bed went beyond silanol $pK_a$, the surface silanol groups were rendered negatively charged and DNA molecules began to desorb. The pH went back to 7.0 after the removal of the voltage and the replacement with fresh buffer. When the flow rate of the DNA solution was increased to 2 μL/min, the whole process was very similar to that with 1 μL/min, as demonstrated in Fig. 4b. However, when we further increased the flow rate to 3 μL/min, due to the dilution caused by the high flow rate, the electrochemistry created much less change in the hydroxide concentration (or pH) than with 1 or 2 μL/min (Fig. 4c). We observed much less pronounced change in the DNA adsorption when the voltage was applied than the other two cases. There were small discontinuities in the increase of fluorescence intensity in Fig. 4 after the first adsorption/desorption cycle. These may possibly be due to fluorescence photobleaching which becomes pronounced when the adsorption is slow.

The switching between adsorptive and desorptive states was also observed when a pH 4.0 buffer was used with a flow rate of 0.3 μL/min (Fig. 5a). The response to the applied voltage in pH 4.0 buffer was more rapid and well-defined than in pH 7.0 buffer. The time for DNA adsorption or desorption to finish in response to the electric switch was ~2 min in Fig. 5a. This was expected based on the drastic decrease in DNA adsorption with higher pH in the range of pH 4.0-6.0 (shown in Fig. 2) and the fact that the same amount of hydroxide generated would cause a larger shift in the pH value at pH 4.0 than at pH 7.0. Similar to in pH 7.0 buffer, the increase in the flow rate of the DNA solution from 0.3 to 1 μL/min led to less change in the DNA adsorption upon the electric actuation (Fig. 5b). Our results indicate that the flow rate of

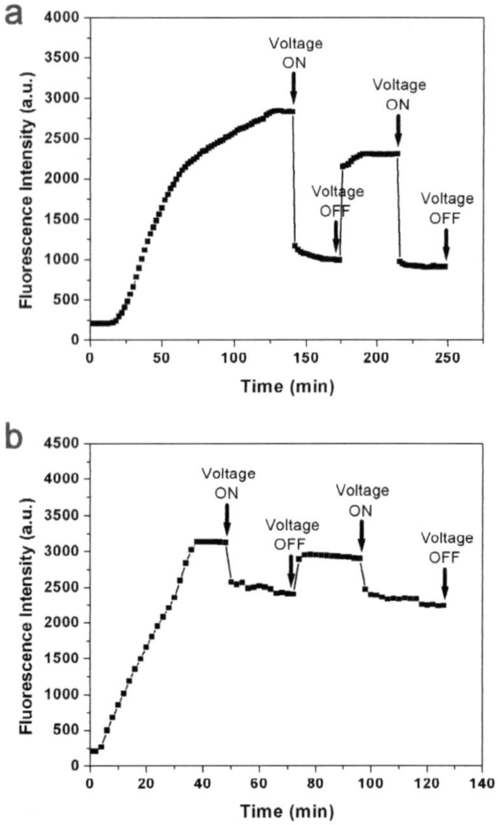


**Fig. 5** The modulation of DNA adsorption on silica beads (10 μm diameter) by electricity in 25 mM acetate buffer solution (pH 4.0) at a flow rate of 0.3 μL/min (a) and 1 μL/min (b). The arrows show the onset and the removal of the constant voltage (22 V). Exposure time for each fluorescent image was 100 ms.

DNA solution has to be compatible with the rate of hydroxide generation and this poses a limitation to the rate of DNA sample being processed. However, we expect that increased electrode surface area may alleviate the problem.

## Conclusions

In summary, we demonstrate the electrical modulation of DNA adsorption and desorption on silica beads in an electrically actuated microfluidic device. Such modulation occurs by varying pH in a packed bed of silica beads through electrolysis of water. We envision that the operational parameters such as the voltage, the flow rate, and the buffer composition and pH can be further optimized for the device to yield optimal performance.

## Acknowledgements

The authors acknowledge support by a cooperative agreement with the ARS-USDA, project number 1935 42000-035, through the Center for Food Safety Engineering at Purdue University, Wallace H. Coulter Foundation, and NSF grant CBET-0747105.

## References

1. Tian, H., Huhmer, A.F. and Landers, J.P. (2000), "Evaluation of silica resins for direct and efficient extraction of DNA from complex biological matrices in a miniaturized format," *Analytical Biochemistry*, **283**, pp. 175-191.
2. Cady, N.C., Stelick, S. and Batt, C.A. (2003), "Nucleic acid purification using microfabricated silicon structures," *Biosensors and Bioelectronics*, **19**, pp. 59-66.
3. Breadmore, M.C., Wolfe, K.A., Arcibal, I.G., Leung, W.K., Dickson, D., Giordano, B.C., Power, M.E., Ferrance, J.P., Feldman, S.H., Norris, P.M. *et al.* (2003), "Microchip-based purification of DNA from biological samples," *Analytical Chemistry*, **75**, pp. 1880-1886.
4. Cao, W., Easley, C.J., Ferrance, J.P. and Landers, J.P. (2006), "Chitosan as a polymer for pH-induced DNA capture in a totally aqueous system," *Analytical Chemistry*, **78**, pp. 7222-7228.

5. Wu, Q., Bienvenue, J.M., Hassan, B.J., Kwok, Y.C., Giordano, B.C., Norris, P.M., Landers, J.P. and Ferrance, J.P. (2006), "Microchip-based macroporous silica sol-gel monolith for efficient isolation of DNA from clinical samples," *Analytical Chemistry*, **78**, pp. 5704-5710.
6. Witek, M.A., Llopis, S.D., Wheatley, A., McCarley, R.L. and Soper, S.A. (2006), "Purification and preconcentration of genomic DNA from whole cell lysates using photoactivated polycarbonate (PPC) microfluidic chips," *Nucleic Acids Research*, **34**, pp. e74.
7. Wen, J., Guillo, C., Ferrance, J.P. and Landers, J.P. (2007), "Microfluidic-based DNA purification in a two-stage, dual-phase microchip containing a reversed-phase and a photopolymerized monolith," *Analytical Chemistry*, **79**, pp. 6135-6142.
8. Woolley, A.T., Hadley, D., Landre, P., deMello, A.J., Mathies, R.A. and Northrup, M.A. (1996), "Functional integration of PCR amplification and capillary electrophoresis in a microfabricated DNA analysis device," *Analytical Chemistry*, **68**, pp. 4081-4086.
9. Burns, M.A., Johnson, B.N., Brahmasandra, S.N., Handique, K., Webster, J.R., Krishnan, M., Sammarco, T.S., Man, P.M., Jones, D., Heldsinger, D. *et al.* (1998), "An integrated nanoliter DNA analysis device," *Science*, **282**, pp. 484-487.
10. Pal, R., Yang, M., Lin, R., Johnson, B.N., Srivastava, N., Razzacki, S.Z., Chomistek, K.J., Heldsinger, D.C., Haque, R.M., Ugaz, V.M. *et al.* (2005), "An integrated microfluidic device for influenza and other genetic analyses," *Lab on a Chip*, **5**, pp. 1024-1032.
11. Poeckh, T., Lopez, S., Fuller, A.O., Solomon, M.J. and Larson, R.G. (2008), "Adsorption and elution characteristics of nucleic acids on silica surfaces and their use in designing a miniaturized purification unit," *Analytical Biochemistry*, **373**, pp. 253-262.
12. Kang, S.H., Shortreed, M.R. and Yeung, E.S. (2001), "Real-time dynamics of single-DNA molecules undergoing adsorption and desorption at liquid-solid interfaces," *Analytical Chemistry*, **73**, pp. 1091-1099.
13. He, Y., Li, H.W. and Yeung, E.S. (2005), "Motion of single DNA molecules at a liquid-solid interface as revealed by variable-angle evanescent-field microscopy," *Journal of Physical Chemistry B*, **109**, pp. 8820-8832.
14. Liu, X., Wu, Z., Nie, H., Liu, Z., He, Y. and Yeung, E.S. (2007), "Single DNA molecules as probes for interrogating silica surfaces after various chemical treatments," *Analytica Chimica Acta*, **602**, pp. 229-235.
15. Isailovic, S., Li, H.W. and Yeung, E.S. (2007), "Adsorption of single DNA molecules at the water/fused-silica interface," *Journal of Chromatography A*, **1150**, pp. 259-266.
16. Nguyen, T.H. and Elimelech, M. (2007), "Plasmid DNA adsorption on silica: kinetics and conformational changes in monovalent and divalent salts," *Biomacromolecules*, **8**, pp. 24-32.
17. Melzak, K.A., Sherwood, C.S., Turner, R.F.B. and Haynes, C.A. (1996), "Driving Forces for DNA Adsorption to Silica in Perchlorate Solutions," *Journal of Colloid and Interface Science*, **181**, pp. 635-644.
18. Li, H.W., Park, H.Y., Porter, M.D. and Yeung, E.S. (2005), "Single DNA molecules as probes of chromatographic surfaces," *Analytical Chemistry*, **77**, pp. 3256-3260.
19. Di Carlo, D., Ionescu-Zanetti, C., Zhang, Y., Hung, P. and Lee, L.P. (2005), "On-chip cell lysis by local hydroxide generation," *Lab on a Chip*, **5**, pp. 171-178.
20. Bard, A.J. and Fox, M.A. (1995), "Artificial Photosynthesis: Solar Splitting of Water to Hydrogen and Oxygen," *Accounts of Chemical Research*, **28**, pp. 141-145.
21. Bao, N., Jagadeesan, B., Bhunia, A.K., Yao, Y. and Lu, C. (2008), "Quantification of bacterial cells based on autofluorescence on a microfluidic platform," *Journal of Chromatography A*, **1181**, pp. 153-158.

22. Wang, H.Y. and Lu, C. (2006), "Electroporation of mammalian cells in a microfluidic channel with geometric variation," *Analytical Chemistry*, **78**, pp. 5158-5164.
23. Kang, S.H., Lee, S. and Yeung, E.S. (2004), "Direct observation of single native DNA molecules in a microchannel by differential interference contrast microscopy," *Analytical Chemistry*, **76**, pp. 4459-4464.

# Programmable Microfluidic Materials with Controlled Shape, Stiffness and Color

Suk Tai Chang, Department of Chemical and Biomolecular Engineering, North Carolina State University, Raleigh, NC

A. Burak Ucar, Department of Chemical and Biomolecular Engineering, North Carolina State University, Raleigh, NC

Sumit Gangwal, Department of Chemical and Biomolecular Engineering, North Carolina State University, Raleigh, NC

Frederick J. Renk, Center for Packaging Innovation, MeadWestvaco, Raleigh, NC

**Orlin D. Velev**, Department of Chemical and Biomolecular Engineering, North Carolina State University, Raleigh, NC

Microfluidic systems for bioanalysis, organic and nanoparticle synthesis have been widely investigated. The potential of microfluidics in other areas of technology, such as making materials with extraordinary adhesion and self-healing properties has only begun to be realized and explored. We will present a class of novel microfluidic materials in the form of flexible sheets that can be solidified on demand by light to acquire specific shapes or that can switch their color. The matrix of these materials is thin sheets of polydimethylsiloxane (PDMS). The microfluidic channel networks embedded in the elastomer are filled with liquid photocurable polymer. The materials formed in this way possess the unique ability to "memorize" and retain user-defined shapes upon illumination. When the microchannel networks are deformed and exposed by UV light, the photoresist inside the channels is solidified and subsequently acts as endoskeleton within the PDMS layer, locking in the programmed shape. The bending and stretching moduli of the materials with solidified endoskeleton increase drastically. The permanent locking in of the shape of light-solidified microfluidic sheets could be used in making instant containers, patches and supports on demand, creating "exoskeletons" for delicate devices, rapid prototyping and multiple other applications. We will also present the results of a technique for making new microfluidic materials that can switch controllably their color and transmittance in the visible and infrared range. Such "chameleon" microfluidic sheets can find applications in smart windows and energy management.

See more of this Session: DNA Analysis in Microfluidic & Nanofluidic Devices
See more of this Group/Topical: Topical 3: 2009 Annual Meeting of the American Electrophoresis Society (AES)

# Structure and Charging Kinetics of Electrical Double Layers at Large Electrode Voltages

Rui Qiao, Mechanical Engineering, Clemson University, Clemson, SC

Clint Cagle, Mechanical Engineering, Clemson University, Clemson, SC

Guang Feng, Mechanical Engineering, Clemson University, Clemson, SC

The structure and charging kinetics of the electrical double layers (EDLs) near polarized electrodes play a key role in many nano-devices such as AC electroosmosis-based nanofluidic pumps. It is thought that the EDLs near electrodes with low voltages and in dilute electrolytes can be described by the classical theories such as the Poisson-Nernst-Planck equations. However, the EDLs near electrodes with large voltages and in concentrated electrolytes are much less understood. Several important works on these problems have been recently reported, e.g., it has been postulated that, at large electrode voltages, the counter-ion concentration near the electrodes may approach the steric limit and modified PNP equations capable of accounting for such effects have been proposed. These works mostly adopt continuum theory approaches and thus have significant advantages in term of generality. However, owing to the complex nature of the EDL, it is challenging to include all aspects of the EDL in such approaches.

In this presentation, we report molecular dynamics (MD) simulation of the EDLs at the interfaces of concentrated NaCl solution and planar electrodes under large voltages. The MD simulation results on the equilibrium structure and charging kinetics of the EDL will be compared against the predictions by the classical Poisson-Nernst-Planck model and the linear RC model.

See more of this Session: Nanoscale Electrokinetics
See more of this Group/Topical: Topical 3: 2009 Annual Meeting of the American Electrophoresis Society (AES)

# DNA Transport Characteristics in Focused Beam-Milled Nanofluidic Devices

Laurent D. Menard Jr.,Chemistry, University of North Carolina at Chapel Hill, Chapel Hill, NC

Chad E. Mair, Chemistry, University of North Carolina at Chapel Hill, Chapel Hill, NC

April L. Holland, Chemistry, University of North Carolina at Chapel Hill, Chapel Hill, NC

Aleksandr Zhushma, Chemistry, University of North Carolina at Chapel Hill, Chapel Hill, NC

Tamoore A. Arshad, Chemistry, University of North Carolina at Chapel Hill, Chapel Hill, NC

J. Michael Ramsey, Chemistry, University of North Carolina at Chapel Hill, Chapel Hill, NC

Fundamental studies of the electrokinetically-driven transport of DNA through nanopores and nanochannels with critical dimensions ranging from 5 to 100 nm are described. Such nanofluidic devices, allowing the spatial confinement of polynucleotides, are of growing importance in fundamental studies of polymer physics and in bioanalytical applications. Nanopores have been demonstrated as sensors for DNA in electrolyte solutions. Single molecules of DNA are eletrokinetically driven through the pore and the standing ionic current is perturbed. The resulting signal is dependent upon the size of the DNA, its associated charge, secondary and tertiary structure, and DNA/pore interactions. This field was launched by the seminal work of Kasianowicz et al. who demonstrated the transport of single-stranded DNA through a α-hemolysin pore [1]. Present investigations continue along these lines using both biological and solid-state nanopores. The differentiation of nucleotides by analyzing changes in the ionic current blockade amplitude has been demonstrated for blocks of repeating nucleotides and work continues to refine the technique to achieve single base resolution [2]. Additionally, solid state nanopores have been proposed as critical components of more complex devices that will measure changes in capacitance as DNA translocates through the pore or a tunneling current across the bases as the molecule passes opposed nanoelectrodes.

Nanochannels are distinguished from nanopores in that a long segment of the polynucleotide (relative to the total contour length of the molecule) is confined during transport. This has several important implications on the characteristics of transport and on the potential utility of these fluidic platforms. If the critical dimensions of the nanochannel are less than the radius of gyration of the molecule in free solution then transport through the channel requires the deformation of the molecule, achieved by extension coaxial to the channel length. If the channel dimensions are greater than the persistence length of DNA (ca. 50 nm for double-stranded DNA in high ionic strength solutions) then this extension is modeled as a series of self-avoiding blobs with diameter constrained by the channel critical dimension

[3]. As the channel size decreases below the persistence length, extension of the molecule in a "single-file" orientation is expected [3]. Nanochannels have been used to separate different sized DNA molecules, size DNA on a single-molecule basis, and map restriction enzyme sites [4-5]. In such experiments, both enthalpic effects (e.g. polynucleotide/wall interactions) and entropic effects due to the decrease in conformational degrees of freedom are of critical importance.

We report the fabrication of nanopores in thin solid-state membranes using either focused electron beam (FEB) or focused ion beam (FIB) milling. In the case of the former, nanopores as small as 1 nm can be fabricated in 10 to 60-nm thick silicon nitride membranes. With FIB-milling, larger nanopores with diameters of 20 nm or larger can be quickly milled in such membranes. We have also developed FIB-milling techniques to mill nanochannels with critical dimensions approaching 5 nm in insulating substrates of SiO2. The nanochannels are interfaced to microfluidic channels formed by photolithography and wet chemical etching and the open trenches are subsequently covered using fusion bonding. The flexibility of FIB milling has allowed us to mill channels that are hundreds of microns down to ca. 50 nm in length, the latter resulting in features analogous to pores in thin membranes. These platforms have allowed fundamental studies of DNA transport in confined environments.

In one study, arrays of nanochannels approximately 50 μm in length were milled across a junction, connecting parallel microfluidic channels. These nanochannels have critical dimensions of 100, 50, and 25 nm (width and depth, with unity aspect ratio). Devices were prepared by first fabricating microfluidic channels on a 25 x 25 $mm^2$ quartz substrate using standard photolithography and wet etching techniques. The nanochannels were then FIB milled through a thin conductive chromium film using a FEI Helios NanoLab dual-beam FIB. The chromium was removed and the device sealed with a quartz coverslip using fusion bonding. Solutions were added to the microchannels through vias that were powder blasted through the substrate prior to FIB milling. Translocation measurements were made in TBE buffer solutions of varying concentration. λ-phage DNA was stained with YOYO®-1 dye at a ratio of 5:1 base pairs:dye molecule and diluted to a concentration of 0.25-1 ng/μL in buffer. Translocations were driven by the application of voltages of 1-20 V across the nanochannels (typical field strengths of 200-4000 V $cm^{-1}$) and were imaged through a 100x plan apochromatic oil immersion objective using an electron multiplying CCD camera with frame rates up to 400 $s^{-1}$. The threshold field strength necessary to overcome the entropic barrier and drive translocations was experimentally determined. The recorded images were analyzed to determine the DNA extension length and dynamic behavior during translocation, the electrophoretic mobility of the DNA, and the velocity of DNA threading into the nanochannels. It was observed that the electrophoretic mobility decreased as the nanochannel critical dimensions decreased, a result commensurate with the increasing channel surface area inducing frictional drag on the DNA. Electrophoretic mobility was found to be independent of DNA size and constant over a range of applied voltages. This latter result differs from results reported for "nanoslits" (channels with micron-scale widths and nanoscale depths) prepared using reactive ion etching [6]. The extension lengths of DNA molecules were found to decrease as the molecules translocated. The proposed model for this effect is one in which the DNA is stretched against the globular portion in the microchannel as it is threaded into the nanochannel and is elastically restored toward a steady-state length once the DNA is fully in the nanochannel. The translocation events are sufficiently short (tens of ms) that slower equilibrations due to entropic restoring forces are not observed

[7].

The translocation of double-stranded DNA through nanopores was also investigated. In these experiments, a voltage across the nanopore and the resulting ionic current were controlled and monitored using a patch clamp amplifier (Axon Instruments Axopatch 200B). DNA solutions were introduced on one side of the pore-containing membrane and electrophoretically driven through the pore to the other side. The localized functionalization of the nanopore in which the pore surface was modified with one moiety while the membrane surface displayed a different surface chemistry was investigated as a means to control DNA/pore interactions. This serves not only to modify mobilities but also to tune the entropic and enthalpic contributions to the barrier to DNA threading, which may further facilitate the discrimination of analytes. This surface functionalization was achieved using well-known and flexible silane chemistry and its localization to the pore interior was characterized using electron energy loss spectrometry (EELS). It was found that by functionalizing the silicon nitride walls of the pore interior with trimethylchlorosilane that the translocation of DNA could be slowed by over 50%. These modifications are also easily applied to long nanochannels milled in SiO2 and comparable alteration of the DNA mobility was demonstrated.

The unifying theme in these and similar experiments enabled by these nanofluidic platforms is the desire to obtain fundamental insights into the behavior of polynucleotide transport in confined environments. There is a practical motivation toward pores and channels with restrictions that are equal to the molecular width of the analyte of interest. In the case of the strategies proposed for the physical sequencing of single DNA molecules (ionic current, capacitance, tunneling current) this condition results in a maximum in signal-to-noise and minimizes the conformational dynamics of the molecule that limit spatial resolution. This work describes efforts to decrease the critical dimensions of nanofluidic pores and channels used for single molecule measurements, modify their surface chemistry, and measure the resulting transport characteristics of polynucleotides.

[1] J.J. Kasianowicz, E. Brandin, D. Branton and D.W. Deamer, Characterization of individual polynucleotide molecules using a membrane channel, Proc. Natl. Acad. Sci. USA 1996, 93, 13770-3.

[2] J. J. Kasianowicz, S. E. Henrickson, H. H. Weetall, B. Robertson, Simultaneous multianalyte detection with a nanometer-scale pore, Anal. Chem. 2001, 73, 2268-72.

[3] W. Reisner, K. J. Morton, R. Riehn, Y. M. Wang, Z. Yu, M. Rosen, J. C. Sturm, S. Y. Chou, E. Frey and R. H. Austin, Statics and dynamics of single DNA molecules confined in nanochannels. Phys. Rev. Lett. 2005, 94, 196101-1-4.

[4] F. Xu and Y. Baba, Polymer solutions and entropic-based systems for double-stranded DNA capillary electrophoresis and microchip electrophoresis, Electrophoresis 2004, 25, 2332-45.

[5] K. Jo, D. M. Dhingra, T. Odijk, J. J. de Pablo, M. D. Graham, R. Runnheim, D. Forrest and D. C. Schwartz, A single-molecule barcoding system using nanoslits for DNA analysis, Proc. Natl. Acad. Sci.

USA 2007, 104, 2673-8.

[6] G. B. Salieb-Beugelaar, J. Teapal, J. van Nieuwkasteele, D. Wijnperlé, J. O. Tegenfeldt, F. Lisdat, A. van den Berg and J. C. T. Eijkel, Field-dependent DNA mobility in 20 nm high nanoslits, Nano Lett. 2008, 8, 1785-90.

[7] J. T. Mannion, C. H. Reccius, J. D. Cross and H. G. Craighead, Conformational analysis of single DNA molecules undergoing entropically induced motion in nanochannels, Biophys. J. 2006, 90, 4538-45.

See more of this Session: Nanoscale Electrokinetics
See more of this Group/Topical: Topical 3: 2009 Annual Meeting of the American Electrophoresis Society (AES)

# Developing a Rapid, Specific and Sensitive Molecular Sensing Kit: An Electrokinetic Nanocolloid Platform

**Satyajyoti Senapati**, Department of Chemical and Biomolecular Engineering, University of Notre Dame, Notre Dame, IN

Sagnik Basuray, Department of Chemical and Biomolecular Engineering, University of Notre Dame, Notre Dame, IN

I-Fang Cheng, Institute of Nanotechnology and Microsystem Engineering / Institute of Biomedical Engineering, National Cheng Kung University, Tainan, Taiwan

Hsueh-Chia Chang, Department of Chemical and Biomolecular Engineering, University of Notre Dame, Notre Dame, IN

The development of highly sensitive detection platform for genetic and peptide identification has attracted great attention to a wide variety of fields such as genomics, proteomics, clinical diagnosis etc.[1] Conventional lab based technique mostly employs optical detection methods, such as in microarray analysis and real-time PCR, which is highly sensitive and specific.[2] However, these techniques require qualified professionals and elaborate fluorescent tagging/sensing equipment/reagents, which severely limit their portability. Furthermore, DNA hybridization reactions in microarray analyses are slow (>hours) due to the diffusion-limiting binding kinetics, making the technique difficult for point-of-need and high-throughput applications.[3] Non-specific binding and interference are also key challenges for protein and DNA microarrays with heterogeneous samples. Thus, there is a need to develop extremely sensitive, rapid, specific, portable and robust DNA/RNA/peptide detection platforms for applications ranging from genomics and proteomics to pathogen identification.[4] We present a new detection platform that can meet these stringent challenges. The platform adopts an open-flow microfluidic format that can concentrate a small number of target molecules in a 100 microliter sample in 10 minutes, thus reducing the transport resistance. The concentration technique involves binding onto probe-functionalized nanobeads whose induced electric dipole reverses direction upon docking at a particular AC field frequency. This dipole reversal upon molecular docking allows us trap the unhybridized beads in an on-chip DEP (dielectrophoretic) trap where the target molecules from the open-flow are trapped. The AC electric field does not impart a force on the flowing liquid and there is hence little hydrodynamic resistance like membrane filters. Moreover, the hybridized beads are released individually and can be trapped and quantified downstream with a nanocolloid impedance sensor. Alternatively, the reversed nanocolloid dipoles can be detected and quantified with an impedance sensor at the DEP trap. As there are about 60 target molecules per nanocolloid and as the impedance resolution is nearly one nanocolloid, the detection sensitivity obtained by this platform is below pM. Moreover, hybridization in the presence of shear flow is highly specific and can distinguish between three bases mismatches when compared with DNA from targeted species and a closely related congeneric species having mismatches at positions 12, 21 and 22 of the 26 base tag sequences (reading from 5' to 3', respectively). This is attributed to large hydrodynamic shear force ($>10^4$ $s^{-1}$) generated by flow that removes any non-specific binding of congener DNA onto the nanobead. The open flow platform also minimizes the masking of the Warburg impedance signal for electron transfer to the nanocolloids by the electrical double layer impedance signature. This sensitive and specific electrokinetic nanocolloid impedance assay for biomolecules hence offers a rapid and portable platform that does not require optical sensing and fluorescent labeling. Genetic detection of specific organisms and pathogens in real-world heterogeneous samples will be demonstrated.

**References**:

[1] E.Palecek, Trends Biotechnol. **22** (2004) 55.

[2] J. H. Watterson, P. A. E. Piunno, U. J. Krull, Anal. Chim. Acta **457** (2002) 29.

[3] J. Kian-Kok Ng, H. Feng, W.T. Liu, Anal. Chim. Acta **582**, (2007) 295.

[4] S. Senapati et al. Biomicrofluidics **3** (2009) 022407.

See more of this Session: Nanoscale Electrokinetics
See more of this Group/Topical: Topical 3: 2009 Annual Meeting of the American Electrophoresis Society (AES)

# A Tunable Voltage-Gated Nanochannel for Electrokinetic Sample Preconcentration

**Aditya S. Khair**, Chemical Engineering, University of California at Santa Barbara, Santa Barbara, CA

X. Trent Huang, Mechanical Engineering, University of California at Santa Barbara, Santa Barbara, CA

Sumita Pennathur, Mechanical Engineering, University of California at Santa Barbara, Santa Barbara, CA

Todd M. Squires, Chemical Engineering, University of California at Santa Barbara, Santa Barbara, CA

In electrokinetic phenomena, the distinction between "nano"-scale and "micro"-scale is an important one, since the former refers specifically to situations involving the overlap of nano-meter sized electrical double layers surrounding adjacent charged surfaces. Indeed, double layer overlap in nanochannels has been utilized by various researchers as a means of affecting ion-selective transport and sample preconcentration of bio-molecules. In these cases, however, the ion selectivity is determined by the native zeta potential of the channel and ionic concentration of the electrolytic solution, quantities that are difficult to vary independently. In this talk, we present theoretical and experimental investigations on a novel voltage-gated nanochannel, wherein metal electrodes are deposited on portions of the top and bottom surfaces of a fused silica nanochannel. An electric field or pressure difference applied axially drives ionic transport along the nanochannel, and by actively controlling the voltage of the gate electrodes we are able to tune the local zeta potential --- and hence ion selectivity --- of the nanochannel. A theoretical analysis is presented to elucidate the degree of ion enrichment (preconcentration) and depletion at opposite sides of the gate, as a function of gate voltage, driving voltage, electrode length (relative to the nanochannel), and ionic concentrations. Experimentally, we visualize and quantify the ionic enrichment and depletion zones using fluorescent tracer molecules and a robust epifluorescent experimental setup.

See more of this Session: Nanoscale Electrokinetics
See more of this Group/Topical: Topical 3: 2009 Annual Meeting of the American Electrophoresis Society (AES)

# An Electrospray Nanocolloid Assay for Rapid Biomarker Detection

**Xinguang Cheng**, Department of Biomolecular & Chemical Engineering, University of Notre Dame, Notre Dame, IN

Satyajyoti Senapati, Department of Chemical and Biomolecular Engineering, University of Notre Dame, Notre Dame, IN

Sagnik Basuray, Chemical and Biomolecular Engineering, University of Notre Dame, Notre Dame, IN

H.-C. Chang, Chemical and Biomolecular Engineering, University of Notre Dame, Notre Dame, IN

The development of fast and highly-sensitive bioassays has attracted considerable attention for a wide variety of application in molecular diagnosis. In this talk, we introduce a new nanocolloid electrospray technology that allows detection of different sized biomolecules in 5 minutes by a rapid and highly sensitive nanocolloid DC electrospray technique. DNA hybridization, antigen-antibody binding and micro-vesicles released from ovarian cancerous cells are shown to be successfully detected by this technique. This is achieved by spraying a mixture, composed of species specific probe (oligomer probe/ antibody/antigen) functionalized nanocolloids with target biomolecules by DC electrospray, which allows nanocolloids and target biomolecules to concentrate at the tip of the Taylor cone of radius smaller than 1 micron. Upon exiting the cone, Rayleigh fission and evaporation endow the aggregates with much larger charge than undocked DNA/antibody and nanocolloids, hence allowing precise separation and deposition of the large aggregates in concentric bands (under atmospheric conditions), whose locations can be predicted from the discrete Laplace harmonics of the Taylor cone. This concentrated field-induced dipolar interaction, which results in the stretching of long target DNA molecules greatly enhance the rate of DNA hybridization. With multiple antibodies functionalized onto the nanocolloids behaving as a network of polyvalent linkers bind the biomarkers to multiple docking sites resulting into aggregates large enough for clear separation in Taylor harmonics. Janus colloids with asymmetric hemispheres are shown to exhibit even higher induced dipoles and to further promote hybridization and aggregation. The aggregates show different connectivity and geometry with the Janus colloid linkers.

See more of this Session: Nanoscale Electrokinetics
See more of this Group/Topical: Topical 3: 2009 Annual Meeting of the American Electrophoresis Society (AES)

# EFFECT of Material Morphology ON Electrophoresis of Bio-Molecules: A Computational-BASED Approach

**Jyothirmai J. Simhadri**, Chemical Engineering, Tennessee Technological University, Cookeville, TN

Mario Oyanader, Chemical Engineering, Tennessee Technological University, Cookeville, TN

Holly Stretz, Chemical Engineering, Tennessee Technological University, Cookeville, TN

Pedro Arce, Chemical Engineering, Tennessee Technological University, Cookeville, TN

Electrophoresis in polymer hydrogels with nanometer-scale pore structure is widely used for the separation and purification of biological macromolecules. In gel-electrophoresis, the internal morphology of the gel also plays an important role in improving the separation. Tuning the nanometer-scale pore structure of the gel either by templating or by adding nanoparticles to improve separations has been the current area of focus in our research group. Moreover, analysis of the effects of the nature of the pore alignment, pore length and diameters on the transport of macromolecules is an important aspect to be studied either analytically or computationally as shown by previous efforts (Trinh et al, 1999; Hidalgo et. al, 2007). In this research, we propose to computationally analyze different pore models (associated with gel materials) and study the effect of geometry on the transport properties in this anisotropic-like media. Results will illustrate, for example, the role of pore-scale in conjunction with electrical fields and identify potentially usual architectures for an optimal separation.

References:

1. Trinh, S., B.R. Locke and P.E. Arce, "Convective and Electroconvective Transport in Non-Uniform Channels with Application to Macromolecular Separations," Separation and Purification Technology, 15, 255 (1999).

2. Hidalgo, R., M. A. Oyanader (*), and P. E. Arce, "Dispersive Mixing Effect Caused by Combined Effect of Channel Morphology and Electrophoretic Mobility in Poiseuille Flows." AIChE Annual Meeting, Salt Lake City, 2007. (*) Speaker.

See more of this Session: Nanoscale Electrokinetics
See more of this Group/Topical: Topical 3: 2009 Annual Meeting of the American Electrophoresis Society (AES)

# Microscale Two-Stage Liquid-Liquid Extraction for Purification of Engineered Proteins

Robert J. Meagher, Sandia National Laboratories, Livermore, CA

Yooli K. Light, Biosystems Research and Development, Sandia National Laboratories, Livermore, CA

Anup K. Singh, Biosystems Research and Development, Sandia National Laboratories, Livermore, CA

We have developed a process for micro-preparative purification of nanogram quantities of engineered proteins. An aqueous two-phase system consisting of polyethylene glycol and potassium phosphate is used to perform a laminar flow two phase extraction on a chip. During the first stage of extraction, an aromatic-rich "partition tag" at the N-terminus of a target protein drives partitioning into the PEG-rich phase of the two-phase system. In between stages, a protease (thrombin) is used for site-specific cleavage of the partition tag, which causes a reversal in the partitioning behavior of the target protein, without affecting other proteins in the mixture. A second stage of laminar flow extraction is performed on chip to back-extract the cleaved protein into the salt-rich phase, away from proteins that had co-partitioned to the PEG phase during the first stage of extraction. The process is mild enough to maintain the structure and function of a fluorescent protein. The two-stage process results in improved purity versus a single-stage process previously reported. Two-phase extraction is an alternative to affinity chromatography for very small amounts of protein. A major advantage of the process is the simplicity of the device, which features no moving parts and no chromatography resins or supports.

See more of this Session: Nanoscale Electrokinetics
See more of this Group/Topical: Topical 3: 2009 Annual Meeting of the American Electrophoresis Society (AES)

# Visualizing the Formation of Natural pH Gradients within Microchannels Using Fluorescent Dyes

Aytug Gencoglu, Dave C Swalm School of Chemical Engineering, Mississippi State University, Mississippi State, MS

Adrienne R. Minerick, Dave C. Swalm School of Chemical Engineering, Mississippi State University, Mississippi State, MS

Hydrolysis reactions at electrode surfaces generate H+ ions at the anode and OH- ions at the cathode. In micro- and nano-scale systems, these ions can cause significant pH shifts in the vicinity of the electrodes, and their transport through the system can cause the formation of so-called "natural pH gradients." Natural pH gradients have been exploited for a number of applications, such as capillary isoelectric focusing. Natural pH gradients may also affect microfluidic systems by altering the surface charge of microchannel walls thus impacting electroosmotic flow, or by causing changes in zwitterionic analytes, affecting their electrokinetic mobility. The effects of electric field strength, microchannel length and diameter on the formation of natural pH gradients in fused silica circular microchannels were studied. The pH sensitive fluorescent dye dextran conjugated 5-(and-6)-carboxy SNARF-1 (C.SNARF-1) was used for the ratiometric measurement of the pH both radially and axiometrically within the microchannels.

See more of this Session: Biomems and Microfluidics: Cell and Biomolecule Analysis
See more of this Group/Topical: Topical 3: 2009 Annual Meeting of the American Electrophoresis Society (AES)

# Microscale Isoelectric Fractionation Membranes for Rapid Sample Purification and Enrichment

Greg J. Sommer, Biosystems Research and Development, Sandia National Laboratories, Livermore, CA

Ying-Chih Wang, Biosystems Research and Development, Sandia National Laboratories, Livermore, CA

Anup K. Singh, Biosystems Research and Development, Sandia National Laboratories, Livermore, CA

Anson V. Hatch, Biosystems Research and Development, Sandia National Laboratories, Livermore, CA

Here we present microscale isoelectric fractionation and enrichment (μ IFE) - a novel method for simultaneously separating and concentrating proteins from complex sample matrices. In μ IFE, species are fractionated into zones based on their isoelectric point (pI - the pH at which a protein has no net charge) by electrophoretically driving them through microchannels supporting a serial array of thin (~50μ m-wide) pH-specific polyacrylamide membranes. Membranes are fabricated via photopolymerization of aqueous monomer solutions titrated to precise pH levels with acrylamido buffering compounds. Multiple membranes can be fabricated simultaneously by adopting our previously-developed technique for establishing and photopolymerizing immobilized pH gradients (IPGs) [1]. Rather than polymerizing a continuous IPG, membrane regions are exposed to polymerizing UV light through a photomask. Up to 20 membranes at unique pH values can be implemented in the current device design.

Stacking within each pI zone results in several orders-of-magnitude enrichment of fractionated proteins within minutes. Individual fractions can then be accessed and manipulated for further on-chip processing and analysis. We demonstrate integration of μ IFE with our previously-developed electrophoretic immunoassay platform. Such integration enables isolation of trace analytes and minor isoforms from high-abundance proteins in biological matrices which complicate immuno-analysis and curb limits of detection.

This work provides a novel, ampholyte-free, microchip-based isoelectric focusing mechanism that enables improved sample isolation and accessibility over other on-chip approaches [2]. Integrating μ IFE with sensitive and specific analysis platforms on-chip promises clear quantitative pictures of low abundance proteins and their subtle modifications from miniscule volumes of complex biological samples, and will enhance the limits of detection of next-generation biological diagnostic devices.

[1] G.J. Sommer, A.K. Singh, A.V. Hatch. "On-Chip Isoelectric Focusing Using Photopolymerized Immobilized pH Gradients" Anal. Chem., 2008, 80, 3327-3333.

[2] G.J. Sommer and A.V. Hatch. "Review: Isoelectric Focusing in Microfluidic Devices" Electrophoresis, 2009, 30, 742-757.

See more of this Session: Biomems and Microfluidics: Cell and Biomolecule Analysis
See more of this Group/Topical: Topical 3: 2009 Annual Meeting of the American Electrophoresis Society (AES)

# Rapid and Sensitive Detection of Drug Resistant Tuberculosis Mutations

**Christa N. Hestekin**, Chemical Engineering, University of Arkansas, Fayetteville, AR

Sowmya Krothapalli, Chemical Engineering, University of Arkansas, Fayetteville, AR

Multidrug resistant tuberculosis (TB) is a major health problem in both developed and developing countries. Mutations in TB bacterial genes such as rpoB, inhA and katG lead to TB drug resistance. Identification of mutations in the rpoB, inhA and katG genes of M. tuberculosis in an accurate, rapid, and inexpensive manner are very important for improving human health. Single strand conformation polymorphism (SSCP) is a simple, rapid and inexpensive method for genetic mutation detection. In order for this technique to be clinically feasible, it must be rapid, automated, and sensitive. In this work, we focused on optimizing capillary electrophoresis (CE)-SSCP, we have investigated the effects of important variables including the polymer separation matrix and the capillary wall coating to meet the high resolution requirements for a method to be used in routine clinical practice. It was observed that 4.5% (w/v) PDMA is the optimum polymer concentration including its ability to act as a coating for the capillary walls. We demonstrated the ability of CE-SSCP to detect TB drug resistance single-base mutations in the rpoB and inhA-mabA regulatory region genes in less than one hour, even when mixed with other mycobacteria.

See more of this Session: Biomems and Microfluidics: Cell and Biomolecule Analysis
See more of this Group/Topical: Topical 3: 2009 Annual Meeting of the American Electrophoresis Society (AES)

# Magnifying Differences in Cross Over Frequency of Bacteria Using Solvent Modification and Antibody Tagged Nanoparticles

**Neha Gujarati**, Biological Engineering, University of Missouri-Columbia, Columbia, MO

Jinwang Tan, Biological Engineering, University of Missouri-Columbia, Columbia, MO

Luis Polo-Parada, Medical Pharmacology and Physiology, University of Missouri-Columbia, Columbia, MO

Shramik Sengupta, Biological Engineering, University of Missouri-Columbia, Columbia, MO

Dielectrophoresis (DEP) can potentially serve as a means to rapidly isolate pathogenic bacterial species and strains of interest from a mixture containing other related, often non-pathogenic, species. Two species can be separated from each other easily using DEP if they have different Cross-Over Frequencies (COFs). The COF depends of a bacterial cell in a given solution depends on the properties (conductivity and dielectric constant) of the cell membrane and the cell interior, and on those of the solution in which it is suspended, and thus each species should have a different characteristic cross-over frequency.

In practice, however, it is seen that these crossover frequencies of bacteria can be very close together, thus making their separation using DEP quite difficult and inefficient. We propose a practical solution to this problem by formulating a "DEP buffer" consisting of Mannitol, Amino Hexanoic Acid, and Sodium Chloride in which the differences in COF are magnified. The buffer is hypertonic, has a low conductivity but high dielectric constant, and by magnifying the contribution of the properties of the cell membrane in determining the COF, enables resolution of many different bacterial species like E.coli, Lactobacillus, Pseudomonas, and Salmonella.

Closely related strains (such as E coli O157 and E coli K-12) however, are still not resolvable. We present a way of doing so using antibody-tagged nano-particles to bind to particular strains. Once these particles bind to the bacteria, the COF of the nano-particle bacteria conjugate differs from that of the bacteria alone.

These practical techniques open the door to detecting the presence of target pathogenic bacteria from mixtures containing large numbers of other bacteria

See more of this Session: Biomems and Microfluidics: Cell and Biomolecule Analysis

See more of this Group/Topical: Topical 3: 2009 Annual Meeting of the American Electrophoresis Society (AES)

# Microfluidic Electroporative Flow Cytometry for Studying Single-Cell Biomechanics

**Ning Bao**, Agricultural and Biological Engineering, Purdue University, West Lafayette, IN

Qian Yang, Biomedical Engineering, Purdue University, West Lafayette, IN

Yihong Zhan, Agricultural and Biological Engineering Department, Purdue University, West Lafayette, IN

Chang Lu, Agricultural and Biological Engineering Department, Purdue University, West Lafayette, IN

Biomechanical properties of cells yield important information on the disease state of cells such as transformation and metastasis. Screening of cells based on their biomechanical properties provides rapid tools for label-free diagnosis and staging of cancers. However, existent single cell techniques for measuring biomechanical properties suffer from low throughput (<1 cell/min). This prevents the application of these assays to a large cell population which produces information with statistical significance. In this study, we applied microfluidics-based electroporative flow cytometry (EFC) that combined electroporation with flow cytometry to study deformability of cells at the single cell level with a throughput of ~5 cells/s. The cell swelling during flow-through electroporation was recorded in real time. We believe that the degree of such swelling was indicative of the cell deformability and the cytoskeleton mechanics. Three cell types (MCF-10A, MCF-7 and 12-O-tetradecanoylphorbol-13-acetate (TPA) treated MCF-7) with different malignancy and metastatic potential were tested using our approach. We found that the more malignant and metastatic cell types exhibited more swelling due to higher cell deformability. Furthermore, the disruption of microtubules by colchicine caused substantial change in the EFC results, which confirmed that EFC data strongly reflected the cytoskeletal mechanics. Finally, the cell type with the highest metastatic potential also suffered the most cell death due to the flow-through electroporation treatment, presumably due to the most substantial cell swelling which could irreversibly rupture the membrane. EFC provides a new method for examining single cell biomechanics with high throughput. We believe that this technique will be useful for mechanistic studies of cytoskeleton dynamics and clinical applications such as diagnosis and staging of cancers in general.

See more of this Session: Biomems and Microfluidics: Cell and Biomolecule Analysis
See more of this Group/Topical: Topical 3: 2009 Annual Meeting of the American Electrophoresis Society (AES)

## DIELECTROPHORETIC FRACTIONATION OF A MIXTURE OF BACTERIA AND YEAST

Héctor Moncada-Hernández, Silvia C. López-Andrade, and Blanca H. Lapizco-Encinas*

*Tecnológico de Monterrey, Monterrey, NL, México.*

### Abstract

Insulator-based dielectrophoresis (iDEP) is a novel technique used for the manipulation of particles by employing insulating structures to create an electric field gradient. In this work, trapping, separation and fractionation of *S. cerevisiae* and *E. coli* cells, employing direct current (DC) electric fields through an array of cylindrical insulating posts is presented. Dielectrophoresis (DEP) is defined as the motion of particles resulting from polarization effects induced by non-uniform electric fields. This technique has a great potential for miniaturization and has also been proven as an effective way of manipulating microbes. However, majority of the research studies reported on dielectrophoretic manipulation of microorganisms were performed employing electrode-based DEP and alternating current electric fields. In this work we report the fractionation of a mixture of *E. coli* and S. *cerevisiae cells employing DC-iDEP.* A glass microdevice containing 10 microchannels that were 10.2-mm long, 2-mm-wide and 15-µm-deep was employed. The microchannels used for these experiments contained an array of 32 cylindrical posts straddled by an inlet and outlet reservoirs. The insulating post array consists of four rows of eight posts each with a diameter 440-µm and arranged 520-µm center-to-center. A DC electric field was applied across the length of the microchannel to create the non-uniform electric field across the array of insulating posts. Different buffer solutions of controlled pH and conductivity were employed to evaluate the effect of suspending medium properties on the dielectrophoretic response of each cell type to determine the minimum voltage required for DEP trapping and therefore to know the conditions for the separation of the mixture of both types of cells.

### Introduction

Conventional methods employed for the detection of microorganisms, usually require high cell population, which are obtained by employing traditional cell culture processes, since pathogens can be present in very low concentration (usually less than 100 CFU/mL).[1,2] Traditional methods are time consuming and labor expensive for they rely on the ability of microorganisms to multiply to visible colonies. Microanalytical electrokinetic (EK) techniques such as electrophoresis, dielectrophoresis, electrorotation and impedance detection have found numerous applications in different fields. Miniaturization offers attractive important advantages to analytical systems such as: reduced sample and reagent consumption, lower cost, short response time, greater sensitivity, improved resolution and portability [3,4].

One important characteristic of microanalytical EK methods is intact microorganisms can be analyzed. The electrical phenotype of a cell is the collection of its electrical properties, such as net charge and polarizability. Cells having significant differences in their electrical phenotypes can be separated employing EK techniques [5].

Dielectrophoresis (DEP) is an EK technique based on the manipulation of particles due to polarization effects in nonuniform electric fields. DEP is attractive for the microbial analysis since it allows separation and concentration in a single step [6]. This technique was discovered by Pohl in 1951 [7], and its first application with microbes was reported in 1966 [8]. However, due to lack of sophisticated electronic circuits and electrode arrays, DEP did not developed in a significant manner until the 1990's. Since the last decade of last century, there has been a spurt of studies on the dielectrophoretic manipulation of microorganisms.[5] Even though iDEP is still a nascent technique, it has already proven to be an effective manner for microorganism manipulation. Several research groups have employed this approach to successfully concentrate microorganisms.[6,9-15]

In this study the fractionation and concentration of sample of *E. coli* and *S. cerevisiae* cells is presented. A microchannel with an array of cylindrical insulating structures was employed with direct current (DC) electric fields, to immobilize and concentrate bacterial and yeast cells. Dielectrophoretic behavior of both types of microbes was studied, estimating the trapping threshold for each cell type. Finally, the separation of bacteria from yeast cells is demonstrated as a function of cells size and dielectric properties.

**Theory of dielectrophoresis.**

Nonuniform electric fields can be achieved applying either alternating current (AC) or direct current (DC) fields. The dielectrophoretic force exerted on a spherical particle can be defined as:

$$F_{DEP} = 2\pi\varepsilon_m r_p^{\,3}\,\mathrm{Re}(f_{CM})\,\nabla E^2 \qquad (1)$$

where $\varepsilon_m$ is the permittivity of the suspending medium, $r_p$ is the radius of the particle, $E$ is the local electric field, and $\mathrm{Re}(f_{CM})$ is the real part of the Clausius-Mossotti (CM) factor expressed as:

$$f_{CM} = \left[\frac{\varepsilon_p^* - \varepsilon_m^*}{\varepsilon_p^* + 2\varepsilon_m^*}\right] \qquad (2)$$

where $\varepsilon_p^*$ and $\varepsilon_m^*$ are the complex permittivities of the particle and the medium, respectively. The complex permittivity is related to the real permittivity $\varepsilon$ by the expression $\varepsilon^* = \varepsilon - (j\sigma/\omega)$, where $\sigma$ and $\omega$ are the conductivity and angular frequency of the applied electric field respectively; while $j = \sqrt{-1}$. The CM accounts for the particle polarizability, a positive value means that the particle is more polarizable than the suspending medium and will be attracted to the regions of higher field intensity *i.e.* positive DEP. A negative value of CM, means that particles are less polarizable than the medium and will be repelled from the regions of higher field intensity, *i.e.* negative DEP. When a direct current (DC) electric field is used, the CM factor can be expressed in terms of the real conductivities of the particle ($\sigma_p$) and the suspending medium ($\sigma_m$) [16]:

$$f_{CM} = \left[\frac{\sigma_p - \sigma_m}{\sigma_p + 2\sigma_m}\right] \qquad (3)$$

When DC electric fields are employed with iDEP, electrophoresis (EP) and electroosmotic flow (EOF) are also present. Electrokinetics (EK) is the superposition of EP and EOF. For a microchannel with a negative surface charge, such as glass, the electrokinetic velocity ($v_{EK}$,) can be defined as follows [17]:

$$v_{EK} = \mu_{EK}E = (\mu_{EO} - \mu_{EP})E \qquad (4)$$

$$v_{EO} = \mu_{EO}E \qquad (5)$$

$$v_{EP} = \mu_{EP} E \tag{6}$$

where $\mu_{EO}$ and $\mu_{EP}$ are the electroosmotic and electrophoretic mobilities, and $v_{EO}$ and $v_{EP}$ are the electroosmotic and electrophoretic velocities, respectively. In order to achieve trapping of particles, the dielectrophoretic velocity must overcome that of electrokinetics.[17] This is possible only under high electric fields, since the former $v_{EK}$ is linearly related to the field, while $v_{DEP}$ has a relationship of second order with the applied field:

$$v_{DEP} = -\mu_{DEP} \nabla E^2 \tag{7}$$

where $v_{DEP}$ and $\mu_{DEP}$ are the dielectrophoretic velocity and mobility, respectively.

## Experimental

### Bacterial cells

*Escherichia coli* (strain Top10) were cultured in Luria-Bertani Broth (LB) at 37° C in a shaker incubator for 15 h to achieve late log phase to a cell concentration of $6 \times 10^8$ cells/ml, verified by OD measurements at 600 nm. Cells were centrifuged at 7000 rpm for 10 min in a microcentrifuge in order to extract the LB broth and re-suspend in DI water with a vortex mixer. Cells were labeled with Syto® 9 bacterial stain (Molecular Probes, Inc., Eugene, OR) that fluoresce green (excitation/emission 483/503 nm).

### Yeast cells

*Saccharomyces cerevisiae* (strain ATCC #24858) were cultured in YM Broth (YMB) at 30° C in a shaker incubator for 18 h to achieve late log phase to a cell concentration of $6 \times 10^6$ cells/ml verified by OD measurements at 600 nm. Yeast cells were washed and resuspended to a desired volume following the same protocol as for bacteria.

### Buffer solution

DI water with adjusted pH (6 to 8) and conductivity (50 and 200 μS/cm) was employed as suspending medium.

### Experimental setup

A schematic representation of the experimental setup is shown in Figure 1. Experiments were conducted in a glass microfluidic chip consisting of two square Schott D263 glass wafers. The top glass wafer contained drilled reservoirs and bottom glass wafer contained microchannels patterned with wet-etching techniques. The microchannel employed for experimentation was 14-μm deep, 10.16-mm long and 2-mm wide containing an array of cylindrical posts, post diameter 440-μm and were arranged 520-μm center-to-center. a post's center to center distance of 520-μm (Figure 1). The first and last row of posts had "dove-tail" geometry in order to avoid particles collapse against the posts and plug the system. A high voltage sequencer model HVS448 (LabSmith, Livermore, CA USA) was used to apply electric fields with platinum-wire electrodes.

Visualization of dielectrophoretic behavior of microorganism was able employing an inverted epifluorescence video microscope for microfluidics model SVM340 (LabSmith, Livermore, CA USA). Microscope objectives 4X and 10X were used for the experiments. Prior to experimentation, the microchannel was filled with buffer solution. After cells sample of 90 μL was introduced into the inlet reservoir, electrodes were placed at in the inlet and outlet reservoirs in order to apply a direct current electric field across the microchannel. Dielectrophoretic behavior of the cells was recorded in the form of videos and pictures using the microfluidics microscope and the computer.

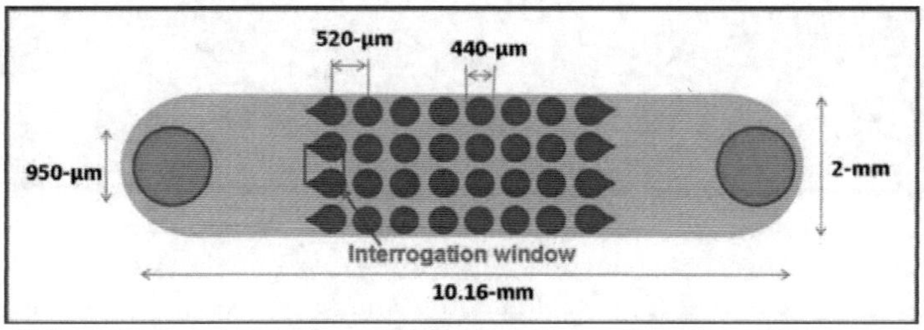


**Figure 1.** Schematic representation of the microchannel employed for experimentation (top view).

### Results and discussion

Microorganisms exhibit negative dielectrophoretic behavior when exposed to DC electric fields.[6] This means that particles will be repelled from the areas of higher field intensity (narrow regions between the cylindrical posts). Additionally, as shown in Eq. (1), the dielectrophoretic force exerted on a particle depends greatly on particle diameter. Since yeast cells have approximate diameter of 5 µm it is expected that they will show a stronger dielectrophoretic response than *E. coli* cells that have a diameter of 1.1 µm.

**Dielectrophoretic trapping of *S. cerevisiae***

A set of experiments with unlabeled yeast cells was carried out. Figure 2 shows some of the results obtained with yeast cells by varying suspending medium and electric field. Figure 2a shows the DEP response obtained with a suspending medium with $\sigma_m$=200 µS/cm and pH=8 applying a field of 200 V/cm. It can be seen that yeast cells are flowing, *i.e.* DEP did not overcame EK. By increasing the field to 500 V with the same suspending medium (Fig 2b), it is possible to observe dielectrophoretic immobilization of yeast cells, with only few cells still flowing. By decreasing the pH of the suspending medium, dielectrophoretic trapping improves since EOF becomes weaker.[17] At an applied field of 200 V/cm, with pH=6, and same $\sigma_m$=200 µS/cm, Figure 2c shows stronger trapping than Figure 2a. Same trend is observed with the results obtained at 500 V/cm, where Figure 2d shows the stronger trapping than Figure 2b, as a result of lowering pH. These results show how dielectrophoretic trapping of cells can be improved by manipulating suspending medium properties.

**Dielectrophoretic trapping of *Escherichia coli***

Analogous as with yeast cells, experiments were carried out to demonstrate the potential of iDEP for the concentration of *Escherichia coli* cells, dyed with Syto 9, these results are reported in Figure 3. Figure 3a shows the dielectrophoretic behavior of *E. coli* cells at 400 V/cm when a suspending medium with $\sigma_m$=200 µS/cm and pH=8 is employed. From this figure it is possible to observe cells flowing with some weak dielectrophoretic trapping. The degree of cell immobilization is increased significantly when the field is increased to 600 V/cm (Fig. 3b). Additionally under the same applied fields of 400 and 600 V/cm, employing same $\sigma_m$=200 µS/cm, but decreasing pH = 6, EOF is weaker, leading to stronger dielectrophoretic trapping shown in Figures 3c and

3d. Significant concentration of *E. coli* cells, demonstrated by increased fluoresce intensity, can be observed from Figure 3d.

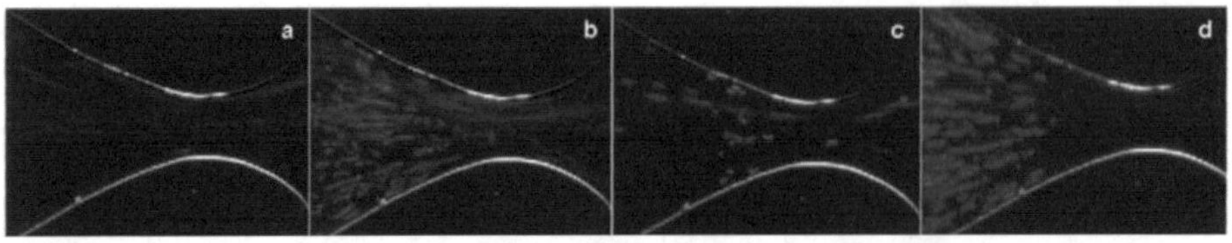


**Figure 2**. DEP performance on *S. cerevisiae* cells employing 10X magnification. (a) and (b) show the results obtained with a suspending medium pH=8 and $\sigma_m$=200 μS/cm, applying electric fields of 200 V/cm and 500 V/cm, respectively. (c) and (d) show the results obtained with a suspending medium pH=6 and $\sigma_m$=200 μS/cm, applying electric fields of 200 V/cm and 500 V/cm, respectively.

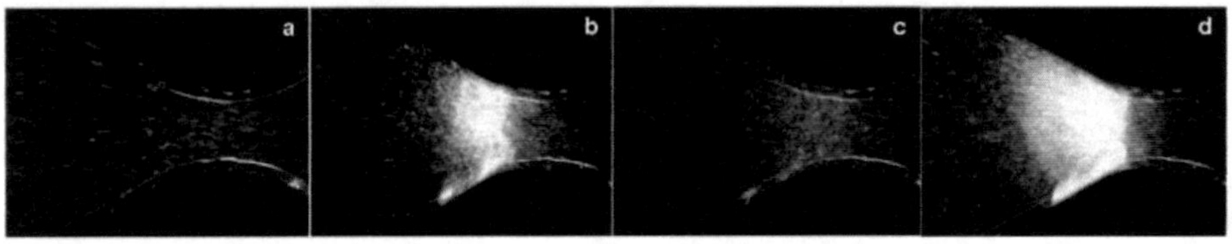


**Figure 3.** Dielectrophoretic response of *E. coli* cells labeled with Syto 9 dye, 10X microscope objective. (a) and (b) show the results obtained with a suspending medium pH=8 and $\sigma_m$=200 μS/cm, applying electric fields of 400 V/cm and 600 V/cm respectively. (c) and (d) show the results obtained with a suspending medium pH=6 and $\sigma_m$=200 μS/cm, applying electric fields of 400 V/cm and 600 V/cm.

As shown in Eq (1), the magnitude of the DEP force exerted on a cell depends greatly on cell size. Due to the differences in size between yeast and *E. coli* cells, it should be possible to achieve dielectrophoretic fractionation of a mixture of these two cell types. In order to explore this possibility, a set of experiments were carried out to evaluate the minimum electric field required for both types of cells in order to achieve dielectrophoretic trapping. These measurements were done by introducing into de system a sample of one cell type and then ramping the voltage up until observing dielectrophoretic trapping. Figure 4 shows, as expected, *E. coli* cells require a higher electric field than yeast cells in order to be dielectrophoretically immobilized. The figure also shows that increasing the suspending medium conductivity improves dielectrophoretic trapping, since it increases the magnitude of the CM factor (Eq. (3)), leading to a lower required field to achieve trapping. Due to this, the applied field required when $\sigma_m$=200 μS/cm are lower than those required when $\sigma_m$=50 μS/cm.

**Separation of a mixture of bacteria and yeast cells**

The results presented above suggest that fractionation of a mixture of these two cells should be possible employing iDEP. A set of experiments was performed with this objective employing a mixture prepared having a yeast to *E. coli* ratio of 1:100, where final concentration was $3x10^6$ and $3x10^8$ for yeast and bacteria, respectively. The results obtained are presented in Figure 5. Figure 5a shows flow of both bacteria and yeast cells across the post array when a field of 200 V/cm was applied. In Figure 5b, one can observe some differences between the responses of both cell types, since yeast cells are trapped meanwhile bacteria cells are flowing through, when the applied field is increased to 500 V/cm. Figure 5b is evidence of dielectrophoretic separation. Finally, figure 5c shows that at a much higher applied electric field (1000 V/cm) both types of cells are dielectrophoretically immobilized, demonstrating the potential of iDEP for cell concentration.

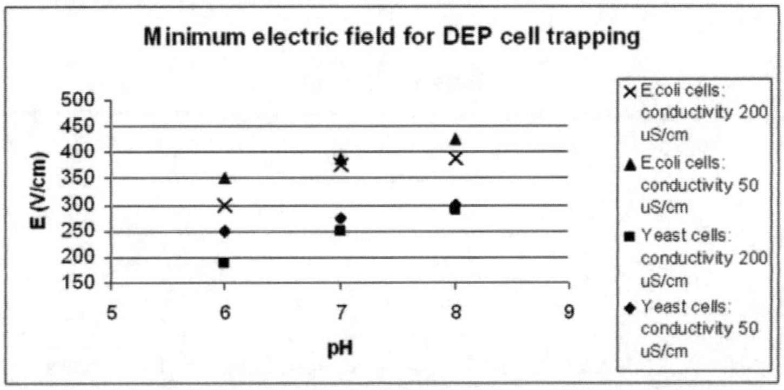


**Figure 4.** Minimum electric field needed for dielectrophoretic trapping of cells as function of suspending medium properties.

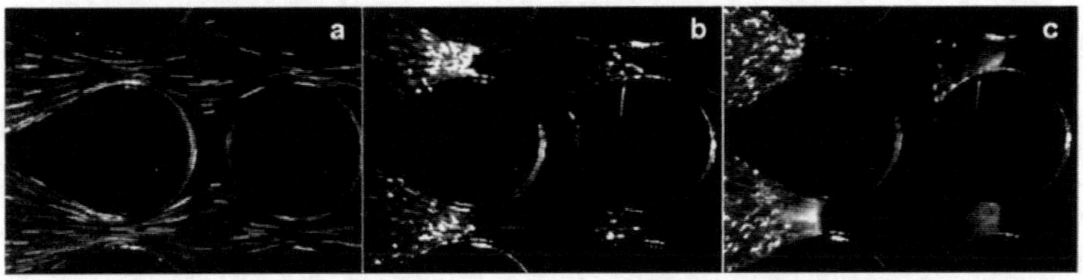


**Figure 5.** DEP performance of a mixture of yeast (large white particles) and bacteria cells (small green particles), 4X microscope objective was employed. (a) show electrokinetic flow of both cells under an electric field of 300 V/cm. (b) DEP trapping of yeast cells and flow of bacteria cells under an electric field of 500 V/cm. (c) DEP trapping of both cells under an electric field of 1000 V/cm.

## Conclusions

The manipulation and separation of microorganisms applying insulator-based dielectrophoresis with direct current electric fields (DC-iDEP) was demonstrated in this work. An electric field gradient was achieved applying direct current along an array of cylindrical insulating posts which constricted the electric field lines across the microchannel creating zones of high intensity in the narrowest section between the posts.

Concentration of each *S. cerevisiae* and *E. coli* cells was achieved at different suspending media conditions in order to determine the optimal conditions for DEP trapping minimizing applied voltage. Dielectrophoretic trapping for both types of cells was demonstrated, and trapping was improved by modifying suspending medium properties. It was demonstrated that *E. coli* cells needed higher fields (~400 V/cm) in order to be dielectrophoretically immobilized, in contrast with *S. cerevisiae* cells which are trapped at relatively low fields (~200 V/cm). Additionally, separation of the mixture of yeast and bacteria cells was successfully achieved, showing that yeast cells were more easily trapped and more repelled from the zone of high intensity electric field compared to bacteria cells.

These results illustrate the potential of iDEP for the selective concentration and separation of microorganisms for several applications. Future work is envisioned to develop an analytical detection method of microorganisms able to quantify cells concentration.


### Acknowledgments

The authors wish to thank the financial support given by the Award *For Women in Science L'óreal-UNESCO-AMC Mexico 2008*, support provided by grant CONACYT-CB-2006-53603 and *Cátedra de Investigación* of *Tecnológico de Monterrey* (CAT 142).


### References


1 Koubova V., Brynda E., Karasova L., Skvor J., Homola J., Dostalek J., Tobiska P., Rosicky J., (2001), "Detection of foodborne pathogens using surface plasmon resonance biosensors," *Sens. Actuator B-Chem.* 74(1), pp.100-105.

2 Naravaneni R., Kaiser J., (2005), "Rapid detection of food borne pathogens by using molecuar techniques," *J. Med. Microbiol.* 54, pp.51-54.

3 Duffy D.C., McDonald J.C., Schueller O.J.A., Whitesides G.M., (1998), "Rapid Prototyping of Microfluidic Systems in Poly(dimethylsiloxane)," *Anal. Chem.* 70(23), pp.4974-4984.

4 Whitesides G.M., (2006), "The origins and the future of microfluidics," *Nature* 442(7101), pp.368-373.

5 Voldman J., (2006), "Electrical forces for microscale cell manipulation," *Ann. Rev. Biomed. Eng.* 8, pp.425-454.

6 Lapizco-Encinas B.H., Simmons B.A., Cummings E.B., Fintschenko Y., (2004), "Dielectrophoretic Concentration and Separation of Live and Dead Bacteria in an Array of Insulators," *Anal. Chem.* 76(6), pp.1571-1579.

7 Pohl H.A., (1951), "The Motion and Precipitation of Suspensoids in Divergent Electric Fields," *J. Appl. Phys.* 22(7), pp.869-871.

8 Pohl H.A., Hawk I., (1966), "Separation of Living and Dead Cells by Dielectrophoresis," *Science* 152(3722), pp.647-649.

9 Lapizco-Encinas B.H., Simmons B.A., Cummings E.B., Fintschenko Y., (2004), "Insulator-Based Dielectrophoresis for the Selective Concentration and Separation of Live Bacteria in Water," *Electrophoresis* 25, pp.1695-1704.

10 Chou C.F., Zenhausern F., (2003), "Electrodeless dielectrophoresis for micro total analysis systems," *IEEE Eng. Med. Biol. Mag.* 22(6), pp.62-67.

11 Zhou G.B., Imamura M., Suehiro J., Hara M., "A dielectrophoretic filter for separation and collection of fine particles suspended in liquid," (2002), in 37th Annual Meeting of the IEEE Industry Applications Society, Pittsburgh, Pennsylvania, vol. 2, pp. 1404-1411.

12 Suehiro J., Zhou G.B., Imamura M., Hara M., (2003), "Dielectrophoretic filter for separation and recovery of biological cells in water," *IEEE Trans. Ind. Appl.* 39(5), pp.1514-1521.

13 Barrett L.M., Skulan A.J., Singh A.K., Cummings E.B., Fiechtner G.J., (2005), "Dielectrophoretic Manipulation of Particles and Cells Using Insulating Ridges in Faceted Prism Microchannels," *Anal. Chem.* 77(21), pp.6798-6804.

14 Pysher M.D., Hayes M.A., (2007), "Electrophoretic and dielectrophoretic field gradient technique for Separating bioparticles," *Anal. Chem.* 79(12), pp.4552-4557.

15 Kang Y., Li D., Kalams S., Eid J., (2008), "DC-Dielectrophoretic separation of biological cells by size," *Biomed. Microdev.* 10(2), pp.243-249.

16 Benguigui L., Lin I.J., (1984), "Phenomenological Aspect of Particle Trapping by Dielectrophoresis," *J. Appl. Phys.* 56, pp.3294-3297.

17 Martínez-López J.I., Moncada-Hernández H., Baylon-Cardiel J.L., Martínez-Chapa S.O., Rito-Palomares M., Lapizco-Encinas B.H., (2009), "Characterization of electrokinetic mobility of microparticles in order to improve dielectrophoretic concentration," *Anal. Bioanal. Chem.* 394(1), pp.293-302.

# A New Strategy for Sample Concentration and Enrichment: Contactless Dielectrophoresis

Hadi Shafiee[1,4], John L.Caldwell[2,4], Michael B.Sano[1,4], Phillip A. Zellner, Masoud Agah[2], and Rafael V.Davalos[1,3,4]

*[1]Engineering Science and Mechanics Department, Virginia Tech, Blacksburg, VA, 24061*
*[2]Bradley Department of Electrical and Computer Engineering, Virginia Tech, Blacksburg VA, 24061*
*[3]School of Biomedical Engineering & Sciences Virginia Tech – Wake Forest University Blacksburg, VA, 24061*
*[4]Bioelectromechanical Systems Laboratory, Institute for Critical Technology and Applied Science (ICTAS), Virginia Tech, VA, 24061*

***Corresponding Author: Rafael V. Davalos Tel: 540-231-1979, E-mail: davalos@vt.edu***

## Abstract

A new strategy to create the non-uniform electric field required for dielectrophoretic (DEP) cell/particle manipulation in microfluidic devices without sample-electrode direct contact is presented. In this method, electrodes are capacitively-coupled to a fluidic channel through dielectric barriers. The application of a high-frequency electric field to these electrodes then induces an electric field in the channel. DEP utilizes the cell/particle electrical signature and size for separation and identification. A microfluidic device with one sample channel and 4 different side electrode channels was designed, fabricated, and tested. The effect of electrode configuration as well as applied voltage and frequency on the gradient of the electric field intensity was investigated numerically and empirically. We observed DEP responses such as pearl chaining and positive and negative DEP forces for human leukemia cancer cells at different voltages and frequencies (100kHz f 300kHz and 50 $V_{rms}$ 150$V_{rms}$). We have also successfully isolated live and dead THP-1 cells in our devices at 200μl/hr. This technique combines the cell manipulation abilities of traditional DEP with the ease of fabrication found in insulator-based technologies.

## Introduction

Biological sample enrichment to detect cells/microparticles is one of the most important steps in many bioengineering applications [1]. Water quality analysis to detect viable pathogenic bacterium [2-6] , the detection of rare circulating tumor cells (CTCs) and their isolation from biological samples to detect cancer at its early stage [7-19] are important examples. Density gradient based centrifugation or membrane filtration [20], fluorescent and magnetic activated cell sorting (F/MACS) [21], cell surface markers [22], and laser tweezers [23] are some of the current enrichment techniques for sample preparation. Each of these techniques relies on different cell properties for separation and has intrinsic advantages and disadvantages. Typically, more sensitive techniques may require prior knowledge of cell-specific markers and antibodies to prepare target cells for analysis.

Dielectrophoresis (DEP), as a technique for sample enrichment, employs the electrical signatures of cells, suspended in biological solutions, and the presence of non-uniform electric fields to separate target cells [24, 25]. The electric field intensity, cell size, medium conductivity and permittivity, as well as the frequency of the applied electric field are significant parameters in a cell's DEP response. Furthermore, it is also believed that the DEP force applied to a cell is strongly dependent on the cell's viability. DEP has been used effectively to isolate target cells including the separation of human leukemia cells from red blood cells in an isotonic solution[26], entrapment of human breast cancer cells from blood[27], separation of human monocytic from peripheral blood mononuclear cells (PBMC)[28], separation of bacteria from blood[29], cancer cells from CD34+ hematopoietic stem cells[30], and to isolate mammalian cells based on their cell-cycle phase[31, 32]. Suehiro et al. utilized dielectrophoretic impedance measurement method to selectively detect viable bacteria [33].

Insulator based dielectrophoresis (iDEP) has been employed to concentrate and separate live and dead bacteria for water analysis [2] and selective trapping of live and dead HeLa cells [34]. In this method, insulator structures in the microfluidic channels disturb the uniform electric field provided by electrodes inserted into the sample channel [35, 36]. Conventional interdigitated electrode DEP micro devices also were used to separate live and heat treated Listeria cells [37]. Huang et al. investigated the difference in the AC electrodynamics of viable and non-viable yeast cells through DEP and electrorotation [38]. A DEP-based microfluidic device for the selective retention of viable cells in culture media with high conductivity was proposed by Docoslis et al. [39].

While many have had success designing and fabricating different DEP and iDEP microdevices to manipulate particles in biological fluids, there are some potential drawbacks of these techniques. The traditional DEP technique suffers from fouling, contamination, bubble formation near integrated electrodes, low throughput, and an expensive and complicated fabrication process [40, 41]. The insulating obstacles employed by iDEP are meant to address these shortcomings and are less susceptible to fouling than integrated electrodes[35]. iDEP's fabrication process is also much less complicated; the insulating obstacles can be patterned while etching the microchannel in one step. This technique has the added benefit of making the process more economical in that mass fabrication can be facilitated through the use of injection molding. Unfortunately, one of the primary drawbacks of an iDEP system is the presence of a high electric field intensity within the highly conductive biological fluid inside the microchannel [40, 42, 43]. The relatively large electrical current flow in this situation causes joule heating and a dramatic temperature increase. The ideal technique would combine iDEP's simple fabrication process and resistance to fouling with DEP's reduced susceptibility to joule heating all-the-while preserving the cell manipulation abilities of both methods.

We have developed an alternative method to provide the spatially non-uniform electric field required for DEP in which electrode/sample contact is avoided [44]. In this method, electrodes are inserted into side channels which are separated from the main channel by thin insulating PDMS barriers. These side channels are filled with a highly-conductive solution such as PBS, and the biological sample is injected into the main channel. The insulating barriers exhibit a capacitive behavior and therefore an electric field can be produced in the main channel by applying an AC field across the barriers. The absence of contact between electrodes and the sample fluid inside the channel prevents bubble formation and avoids any contaminating effects the electrodes may have on the sample. We have designed and fabricated a microfluidic device based on this new technique and we have observed DEP responses in human leukemia, breast, and prostate cancer cells. We have also successfully isolated live leukemia cancer cells from the dead ones without cell lysing.

## METHODS

*Microfabrication*

Deep Reactive Ion Etching (DRIE) was used to etch a <100> silicon master stamp to a depth of 50um. Surface "scalloping" produced during the DRIE process can greatly inhibit the removal of cured PDMS from this stamp. In order to reduce the surface roughness produced during the DRIE process, silicon oxide was grown on the silicon master using thermal oxidation and then removed using HF solvent. The liquid phase PDMS was made by mixing the PDMS monomers and a curing agent in a 10:1 ratio (Sylgrad 184, Dow Corning, USA). Bubbles in the liquid PDMS were removed by exposing the mixture to vacuum for 15 minutes. The bubble-free PDMS liquid was then poured onto the silicon master and cured for 45 min at 100°C. The PDMS portion of the device was then peeled off the mold and fluidic connections to the channels were punched with 15 gauge blunt needles (Howard Electronic Instruments, USA). Cleaned glass microscope slides and the PDMS replica were bonded after treating with oxygen plasma for 40 s at 50 W RF power. A schematic with dimensions and equivalent circuit model of our device is presented in Figure1. The channel depth in this device is 50 μm and the thickness of the PDMS barrier between the side channels and the main channel is 20 μm.

*Cells and buffer*

The live samples of THP-1 human Leukemia monocytes were washed twice and resuspended in our prepared DEP buffer (8.5%sucrose [wt/vol], 0.3% glucose [wt/vol], and 0.725% [wt/vol] RPMI) [45] to achieve 106 cells/ml cell concentration. The electrical conductivity of the buffer was measured with a Mettler Toledo SevenGo pro conductivity meter (Mettler-Toledo, Inc., Columbus, OH) to ensure that its conductivity was 130μS/cm. These cells were observed to be spherical with a diameter of 15.4±2 μm (n=30) when in suspension. The cells suspended in DEP buffer to be killed were first inserted in conical test tubes and were then heated in a 60°C water bath for twelve minutes.

To enable simultaneous observation under fluorescent microscope cells were stained using fluorescent dyes from a LIVE/DEAD® Viability/Cytotoxicity Kit for mammalian cells (Molecular Probes Inc.). Calcein AM, which is enzymatically converted to green fluorescent calcein, was added to the live cell sample at 2μl per ml of cell suspension. Ethidium homodimer-1 (EthD-1) was added to the dead cell sample at 6μl per ml of cell suspension. EthD-1 can only pass through damaged cell membranes and upon nucleic acid-binding produces a red fluorescence. The live and dead suspensions were then mixed together in one conical tube.

*Experimental Set-Up*

The microfluidic devices were placed in a vacuum jar for 30 minutes prior to experiments to reduce problems associated with priming. Pipette tips inserted in the punched holes in PDMS were used as reservoirs to fill the side channels with PBS. Pressure driven flow was provided in the main channel using a microsyringe pump. Once the main channel was primed with the cell suspension the syringe pump was set to 1ml/hr steadily decreasing the flow rate down to 0.02ml/hr (20μL/hr) equivalent to a velocity of ~550μm/sec. This flow rate was maintained for 1 minutes prior to experiments. An inverted light microscope equipped with color camera (Leica DMI 6000B, Leica Microsystems, Bannockburn, IL) was used to monitor the cells flowing through the main channel. High-frequency electric fields were provided by a wideband, high-power amplifier and transformer combination (Amp-Line Corp., Oakland Gardens, NY) and signal generation was accomplished using a function generator (GFG-3015, GW Instek, Taipei, Taiwan).

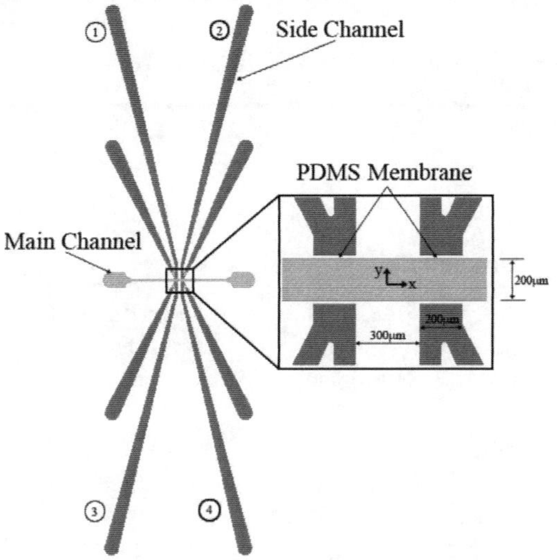


Figure 1. 2D Device Schematic. The channel depth is 50μm.

*Numerical Modeling*

The electric field distribution and its gradient $\nabla E = \nabla(\nabla \phi)$ was modeled numerically in Comsol multi-physics 3.5 using the AC/DC module (Comsol Inc., Burlington, MA, USA). This is done by solving for the potential distribution, $\phi$, using the Laplace equation, $\nabla \cdot (\sigma^* \nabla \phi) = 0$. Where $\sigma^*$ is the complex conductivity of the sub-domains of the microfluidic device. The boundary conditions used are prescribed uniform potentials at the inlet or outlet of the side channels. The electrical conductivity and the relative electrical permittivity of PDMS have been reported as 0.83×10-12 S/m and 2.65 respectively (Sylgrad 184, Dow Corning, USA). The electrical conductivity of PBS and the DEP buffer are 1.4S/m and 130 μS/cm respectively and a relative permittivity of 80. The gradient of the electric field along the center line of the main channel was investigated for different electrode configurations.

## Results

Maximum gradient of the electric field intensity along the center line of the main channel was studied as a function of applied voltage and frequency. The gradient of the electric field intensity increases by increasing the applied voltage up to $500V_{rms}$. However, the gradient of the electric field intensity does not necessarily increase by increasing the frequency; it increases up to 10MHz and then decreases up to 60MHz. This behavior is due to the capacitance effect of the thin PDMS barriers between the side and the main channel and the capacitance effect of the main channel itself.

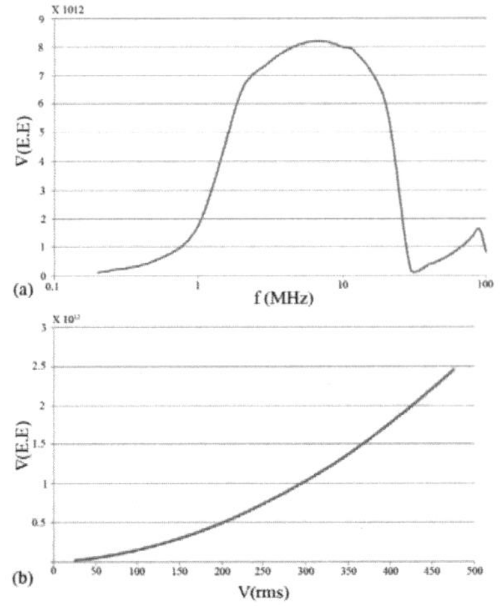


Figure2. Maximum gradient of the electric field along the centerline of the main channel as a function of frequency (a) and voltage (b). (a) $V_1=V_2=50V_{rms}$ at 0.2kHz f 100MHz, $V_3=V_4$=Ground. 50 $V_{rms}$ $V_1=V_2$ 500$V_{rms}$ at f=100kHz, $V_3=V_4$=Ground.

The effect of the electrode configuration in the 4 different side channels on the gradient of the electric field intensity along the centerline of the main channel was also investigated. Four different configurations with the same applied voltage and frequency were studied as shown in Figure 3. The effect of having the charged electrodes inserted in channels 1 and 2 and the ground electrodes in channels 3 and 4 (case 1) is almost similar to the configuration with charged electrodes in channels 1 and 4 and ground electrodes in channel 2 and 3 (case 3). This also repeats for the cases with charged electrodes are in channels 1, 2, and 4 (case 2) and the ground electrode in channel 3 and charged electrode in channel 1 and ground electrode in channel 2 (case 4).

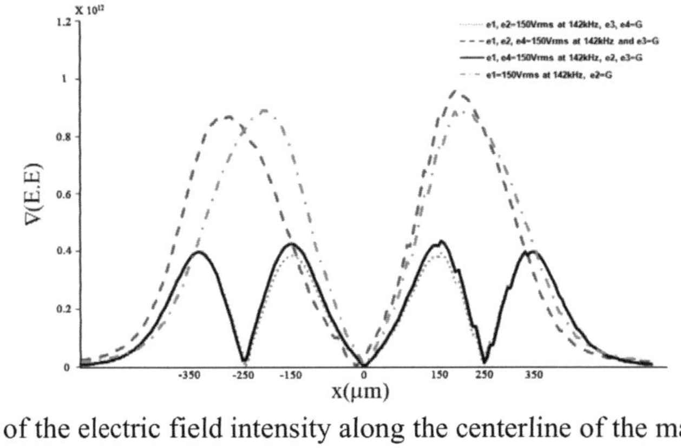


Figure 3. The gradient of the electric field intensity along the centerline of the main channel for different electrode configuration. The electrodes are charged with 150Vrms and 142kHz in the side channels in all different configurations. Case1: charged electrodes are in channels 1 & 2 and ground electrodes are in channels 3 & 4, Case 2: charged electrodes are in channels 1, 2 & 4 and ground electrodes are in channels 3, Case 3: charged electrodes are in channels 1 & 4 and ground electrodes are in channels 2 & 3, Case 4: charged electrodes are in channel 1 and ground electrodes are in channel 2.

These numerical results indicate that the electrode configuration has a substantial effect on the gradient of the electric field intensity and DEP cell manipulation. The electric field gradient intensity in cases 2 and 4 is stronger than cases 1 and 3. Furthermore, one can change the cell/particle manipulation strategy by changing the electrode configurations. The configuration used in case 4 (electrodes in just one side of the main channel) can deflect the target cell/particle trajectory in the main channel such that it ends up in a specific reservoir.

The live THP-1 cells were observed to be trapped efficiently due to positive DEP force at $V_1$= $V_2$=100$V_{rms}$ at 152kHz, $V_3$= $V_4$=Ground (Fig.4). Particles parallel to the electric field attract each other because of the dipole-dipole interaction, resulting in pearl-chaining of the trapped cells in the direction of the electric field in the microfluidic channel[24, 25, 46]. We have also observed cell chain formation in our experiments with an applied AC electric filed.

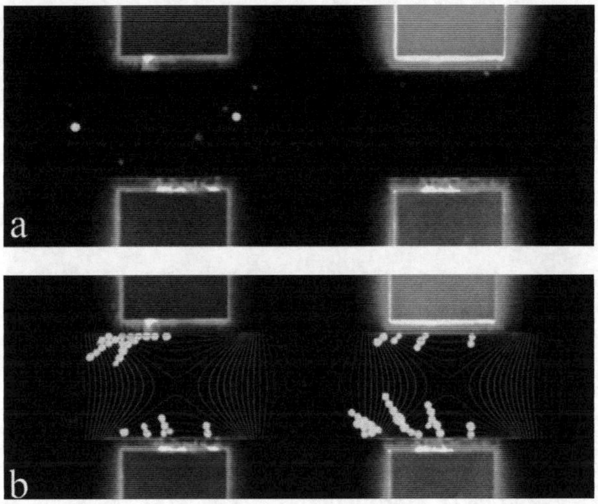


Figure 4: Experimental results: (a) Dead (red) and live (green) THP-1 cells are moving from right to left due to pressure driven flow without applying electric field. (b) 30 seconds after applying the electric field (V1=V2=100Vrms at 152kHz and V3=V4=Ground). The live (green) cells were trapped due to positive DEP, but the dead (red) cells pass by the trapping area. The contours in figure b are superimposed from the numerical results.

The major advantages of contactless dielectrophoresis over other methods of cell sorting include a lack of extensive sample preparation (no antibody labeling, one needs only prepare a single cell preparation) and the speed of isolation (minutes versus hours from the time of sample acquisition). This rapid turnaround is extremely important when evaluating gene expression. We believe this new technique has many attractive features when compared to traditional DEP and insulator based dielectrophoresis (iDEP). Reduced joule heating, no direct electrode contact resulting in less potential for contamination (important for animal studies), simplified inexpensive fabrication process are the other noticeable advantageous of this new technique because the induced electric field is not as intense as comparable methods and is focused just at the trapping zones, we theorize the joule heating within the main microfluidic channel is insignificant. This could mitigate the stability and robustness issues encountered with conventional iDEP, due to the conductivity distribution's strong dependence on temperature. Furthermore, challenges associated with cell lysing due to high temperatures or irreversible electroporation due to high field strengths are overcome with our new design approach under these conditions. We have demonstrated a new technique for inducing electric fields in microfluidic channels in order to create a dielectrophoretic force. Our method relies on the application of a high-frequency AC electric signal to electrodes that are capacitively coupled to a microfluidic channel. In our device, the geometry of the electrode channels create the spatial non-uniformities in the electric field required for DEP. This result highlights the ability of our technique to differentiate cells by their intrinsic electrical properties. We believe this technique may help overcome many of the challenges faced with traditional DEP and iDEP.

References:


1. Morgan, H. and N.G. Green, *AC Electrokinetics: Colloids and Nanoparticles*. 2003, Hertfordshire, England: Research Studies Press LTD. 324.
2. Lapizco-Encinas, B.H., B.A. Simmons, E.B. Cummings and Y. Fintschenko, *Dielectrophoretic concentration and separation of live and dead bacteria in an array of insulators.* Anal Chem, 2004. **76**(6): p. 1571-9.
3. Armstrong, D., J. Schneiderheinze, J. Kullman and L. He, *Rapid CE microbial assays for consumer products that contain active bacteria.* FEMS Microbiology Letters, 2001. **194**(1): p. 33-37.
4. Armstrong, D., G. Schulte, J. Schneiderheinze and D. Westenberg, *Separating microbes in the manner of molecules. 1. Capillary electrokinetic approaches.* Analytical Chemistry, 1999. **71**(24): p. 5465-5469.
5. Girod, M. and D.W. Armstrong, *Monitoring the migration behavior of living microorganisms in capillary electrophoresis using laser-induced fluorescence detection with a charge-coupled device imaging system.* Electrophoresis, 2002. **23**(13): p. 2048-56.
6. Cabrera, C.R. and P. Yager, *Continuous concentration of bacteria in a microfluidic flow cell using electrokinetic techniques.* Electrophoresis, 2001. **22**(2): p. 355-62.
7. Becker, F.F., X.-B. Wang, Y. Huang, R. Pethig, J. Vykoukal and P.R.C. Gascoyne, *The removal of human leukaemia cells from blood using interdigitated microelectrodes.* J. Phys. D: Appl. Phys., 1994. **27**: p. 2659-2662.
8. Gascoyne, P.R.C., X.-B. Wang, Y. Huang and F.F. Becker, *Dielectrophoretic Separation of Cancer Cells from Blood.* IEEE Trans. Industry Applications, 1997. **33**(3): p. 670-678.
9. Huang, Y., S. Joo, M. Duhon, M. Heller, B. Wallace and X. Xu, *Dielectrophoretic cell separation and gene expression profiling on microelectronic chip arrays.* Anal Chem, 2002. **74**(14): p. 3362-71.
10. Cheng, J., E.L. Sheldon, L. Wu, M.J. Heller and J.P. O'Connell, *Isolation of cultured cervical carcinoma cells mixed with peripheral blood cells on a bioelectronic chip.* Anal Chem, 1998. **70**(11): p. 2321-6.
11. Altomare, L., M. Borgatti, G. Medoro, N. Manaresi, M. Tartagni, R. Guerrieri and R. Gambari, *Levitation and movement of human tumor cells using a printed circuit board device based on software-controlled dielectrophoresis.* Biotechnol Bioeng, 2003. **82**(4): p. 474-9.
12. Markx, G.H., M.S. Talary and R. Pethig, *Separation of viable and nonviable yeast using dielectrophoresis.* Journal of Biotechnology, 1994. **32**(1): p. 29-37.

13. Das, C.M., F. Becker, S. Vernon, J. Noshari, C. Joyce and P.R. Gascoyne, *Dielectrophoretic segregation of different human cell types on microscope slides.* Anal Chem, 2005. **77**(9): p. 2708-19.
14. Gascoyne, P.R., J. Noshari, T.J. Anderson and F.F. Becker, *Isolation of rare cells from cell mixtures by dielectrophoresis.* Electrophoresis, 2009. **30**(8): p. 1388-98.
15. Apostolaki, S., M. Perraki, A. Pallis, V. Bozionelou, S. Agelaki, P. Kanellou, A. Kotsakis, E. Politaki, K. Kalbakis, A. Kalykaki, L. Vamvakas, V. Georgoulias, and D. Mavroudis, *Circulating HER2 mRNA-positive cells in the peripheral blood of patients with stage I and II breast cancer after the administration of adjuvant chemotherapy: evaluation of their clinical relevance.* Ann Oncol, 2007. **18**(5): p. 851-8.
16. Cristofanilli, M., *The "microscopic" revolution in breast carcinoma.* Cancer, 2005. **103**(5): p. 877-80.
17. Fizazi, K., L. Morat, L. Chauveinc, D. Prapotnich, R. De Crevoisier, B. Escudier, X. Cathelineau, F. Rozet, G. Vallancien, L. Sabatier, and J.C. Soria, *High detection rate of circulating tumor cells in blood of patients with prostate cancer using telomerase activity.* Ann Oncol, 2007. **18**(3): p. 518-21.
18. Hayes, D.F., M. Cristofanilli, G.T. Budd, M.J. Ellis, A. Stopeck, M.C. Miller, J. Matera, W.J. Allard, G.V. Doyle and L.W. Terstappen, *Circulating tumor cells at each follow-up time point during therapy of metastatic breast cancer patients predict progression-free and overall survival.* Clin Cancer Res, 2006. **12**(14 Pt 1): p. 4218-24.
19. Naoe, M., Y. Ogawa, J. Morita, K. Omori, K. Takeshita, T. Shichijyo, T. Okumura, A. Igarashi, A. Yanaihara, S. Iwamoto, T. Fukagai, A. Miyazaki, and H. Yoshida, *Detection of circulating urothelial cancer cells in the blood using the CellSearch System.* Cancer, 2007. **109**(7): p. 1439-45.
20. Giddings, J.C., *Field-flow fractionation: analysis of macromolecular, colloidal, and particulate materials.* Science, 1993. **260**(5113): p. 1456-65.
21. Miltenyi, S., W. Muller, W. Weichel and A. Radbruch, *High gradient magnetic cell separation with MACS.* Cytometry, 1990. **11**(2): p. 231-8.
22. Fu, A.Y., C. Spence, A. Scherer, F.H. Arnold and S.R. Quake, *A microfabricated fluorescence-activated cell sorter.* Nat Biotechnol, 1999. **17**(11): p. 1109-11.
23. Ashkin, A., J.M. Dziedzic and T. Yamane, *Optical trapping and manipulation of single cells using infrared laser beams.* Nature, 1987. **330**(6150): p. 769-71.
24. Pohl, H.A., *Dielectrophoresis: The Behavior of Neutral Matter in Nonuniform Electric Fields.* Cambridge University Press, Cambridge, U.K., 1978.
25. Pohl, H.A., *Some Effects of Nonuniform Fields on Dielectrics.* J. Appl. Phys. , 1958. **29**.
26. Becker, F., X.-B. Wang, Y. Huang, R. Pethig, J. Vykoukal and P.R. Gascoyne, *The removal of human leukaemia cells from blood using interdigitated microelectrodes.* J. Phys. D: Appl. Phys, 1994. **27**: p. 2659-2662.
27. Gascoyne, P., X.-B. Wang, Y. Huang and F. Becker, *Dielectrophoretic Separation of Cancer Cells from Blood.* IEEE Trans. Industry Applications, 1997. **33**: p. 670-678.
28. Huang, Y., S. Joo, M. Duhon, M. Heller, B. Wallace and X. Xu, *Dielectrophoretic cell separation and gene expression profiling on microelectronic chip arrays.* Anal Chem, 2002. **74**: p. 3362-3371.
29. Cheng, J., E. Sheldon, L. Wu, A. Uribe, L. Gerrue, J. Carrino, M. Heller and J. O'Connell, *Preparation and hybridization analysis of DNA/RNA from E. coli on microfabricated bioelectronic chips.* Nat Biotechnol, 1998. **16**: p. 541-546.

30. Stephens, M., P. Talalay, R. Pethig, A. Burnett and K. Mills, *The dielectrophoresis enrichment of CD34+ cells from peripheral blood stem cell harvests.* Bone Marrow TRansplant, 1996. **18**: p. 777-782.
31. Huang, Y., X. Wang, F. Becker and P. GAscoyne, *Introducing dielectrophoresis as a new force field for field-flow fractionation.* Biophys J, 1997. **73**: p. 1118-1129.
32. Kim, U.-J., C. Shu, K. Dane, P. Daugherty, J. Wang and H. Soh, *Selection of mammalian cells based on their cell-cycle phase using dielectrophoresis.* Proc Natl Acad Sci, 2007. **104**: p. 20708-20712.
33. Suehiro, J., R. Hamada, D. Noutomi, M. Shutou and M. Hara, *Selective detection of viable bacteria using dielectrophoretic impedance measurement method.* Journal of Electrostatics, 2003. **57**(2): p. 157-168.
34. Jen, C.P. and T.W. Chen, *Selective trapping of live and dead mammalian cells using insulator-based dielectrophoresis within open-top microstructures.* Biomed Microdevices, 2009. **11**(3): p. 597-607.
35. Cummings, E.B. and A. Singh, *Dielectrophoresis in Microchips Containing Arrays of Insulating Posts: Theoretical and Experimental Results.* Analytical Chemistry, 2003. **75**: p. 4724-4731.
36. Davalos, R.V., G.J. McGraw, T.I. Wallow, A.M. Morales, K.L. Krafcik, Y. Fintschenko, E.B. Cummings and B.A. Simmons, *Performance impact of dynamic surface coatings on polymeric insulator-based dielectrophoretic particle separators.* Anal Bioanal Chem, 2008. **390**(3): p. 847-855.
37. Li, H. and R. Bashir, *Dielectrophoretic separation and manipulation of live and heat-treated cells of Listeria on microfabricated devices with interdigitated electrodes.* Sensors and Actuators B-Chemical, 2002. **86**(2-3): p. 215-221.
38. Huang, Y., R. Holzel, R. Pethig and X.B. Wang, *Differences in the AC electrodynamics of viable and non-viable yeast cells determined through combined dielectrophoresis and electrorotation studies.* Phys Med Biol, 1992. **37**(7): p. 1499-517.
39. Docoslis, A., N. Kalogerakis, L.A. Behie and K.V. Kaler, *A novel dielectrophoresis-based device for the selective retention of viable cells in cell culture media.* Biotechnol Bioeng, 1997. **54**(3): p. 239-50.
40. Steffen Hardt, F.S., *Microfluidic Technologies for Miniaturized Analysis Systems*. Book, ed. S.D. Senturia. 2007: Springer.
41. Hughes, M.P., *Strategies for dielectrophoretic separation in laboratory-on-a-chip systems.* Electrophoresis, 2002. **23**(16): p. 2569-82.
42. Sabounchi, P., M.P. Kanouff, D.E. Huber, A.E. Harris and B.A. Simmons. *Joule Heating Effects on Insulator-based Dielectrophoresis*. in *Twelfth International Conference on Miniaturized Systems for Chemistry and Life Sciences*. 2008. San Diego, California, USA.
43. Sabounchi, P., A.M. Morales, P. Ponce, L.P. Lee, B.A. Simmons and R.V. Davalos, *Sample concentration and impedance detection on a microfluidic polymer chip.* Biomed Microdevices, 2008. **10**(5): p. 661-70.
44. Shafiee, H., J.L. Caldwell, M.B. Sano and R.V. Davalos, *Contactless dielectrophoresis: a new technique for cell manipulation.* Biomed Microdevices, 2009.
45. Flanagan, L.A., J. Lu, L. Wang, S.A. Marchenko, N.L. Jeon, A.P. Lee and E.S. Monuki, *Unique dielectric properties distinguish stem cells and their differentiated progeny.* Stem Cells, 2008. **26**(3): p. 656-65.
46. Dussaud, A.D., *Particle segregation in suspensions subject to high-gradient ac electric fields.* J. Appl. Phys., 2000. **88**(5463): p. 5463-5473.

# Microfluidic Chamber Array for Continuously-Perfused Cell Culture and Assay

**Edward S. Park**, Chemical and Biomolecular Engineering, Georgia Institute of Technology, Atlanta, GA

Ashley E. Carson, Biomedical Engineering, Georgia Institute of Technology and Emory University, Atlanta, GA

Michael A. DiFeo, Chemical and Biomolecular Engineering, Georgia Institute of Technology, Atlanta, GA

Thomas H. Barker, Biomedical Engineering, Georgia Institute of Technology and Emory University, Atlanta, GA

Hang Lu, Chemical and Biomolecular Engineering, Georgia Institute of Technology, Atlanta, GA

Continuous perfusion during cell culture is desirable for greater control of the microenvironment, particularly when controlling for the effects of autocrine or paracrine signaling. Microfluidic devices are well-suited for continuous perfusion because they offer dramatic savings of reagents due to their small volumes. A variety of microfluidic culture devices has been developed for continuous perfusion, yielding impressive demonstrations of high-throughput data collection. However, these systems typically link culture chambers with shared flow paths, leading to cross-contamination. In addition, the integration of other functions, such as independent addressing (or isolation) of chambers, becomes challenging because of the need for more sophisticated flow networks.

We realize a rectangular chamber array in a PDMS device with three attributes: (i) continuous perfusion; (ii) independently-addressable chambers and subarrays; and (iii) flow paths that forbid cross-chamber contamination. The combination of these attributes is made possible by a bridge-and-underpass architecture, where flow streams travel vertically to pass over (or under) channels and on-chip valves. The array achieves considerable versatility due to subarray, row, column, or single chamber addressing for the following: incubation with adsorbed molecules; perfusion of differing media; seeding or extraction of cells; and assay staining. In addition, culture surfaces are shielded from direct perfusion to minimize shear-induced cell behavior. We have thus far used the device to classify different phenotypes of alveolar epithelial type II (ATII) cells, particularly the extent of epithelial-to-mesenchymal transition (EMT), which is a highly suspected pathway in tissue regeneration, fibrosis, and cancer metastasis. Cells are cultured on combinations of matrix proteins (fibronectin or laminin by row) and soluble signals (TGF-β1 +/- by column) in a 2x4 array with two subarray repeats per chip. We perform cultures up to 4 days, as well as vital and end-point fluorescent assays by subarrays to quantify viability, transdifferentiation (EMT), and F-actin organization. Assay and morphological data are used to tease-out effects of cues driving each phenotype, confirming this as an effective and versatile combinatorial screening platform for trials requiring high microenvironmental control. -11-2009-->

See more of this Session: Biomems and Microfluidics: Cell and Biomolecule Analysis
See more of this Group/Topical: Topical 3: 2009 Annual Meeting of the American Electrophoresis Society (AES)

# Development of a Microfluidics-Based Gel Protein Recovery System

**Jifeng Chen**, Protea Biosciences, Morgantown, WV

Trust Tariro Razunguzwa, Protea Biosciences, Morgantown, WV

April Biddle, Protea Biosciences, Morgantown, WV

Heather Anderson, Protea Biosciences, Morgantown, WV

Dongliang Zhan, Protea Biosciences, Morgantown, WV

Matthew Powell, Protea Biosciences, Morgantown, WV

Reid Asbury, Protea Biosciences, Morgantown, WV

A polymethylmethacrylate (PMMA) microfluidic chip, in which fluid is manipulated to transport protein from a polyacrylamide gel electrophoresis (PAGE) gel piece to a collection reservoir via a microfluidic channel, has been developed. The protein sample is mobilized out of the gel (loaded in a chip access hole) into a low electroosmotic flow capillary electrophoresis (CE) microfluidic channel under the influence of an electric field. Simultaneously, hydrostatic pressure from the filled buffer reservoirs is used to direct the protein sample to a third reservoir, through a field-free channel connected to the electrophoresis channel. Using this novel process of protein transport from a gel sample, proteins from coomassie-stained gels have been transferred into solution in 15-30 min, with good sample recovery, using a run buffer containing an anionic acid labile surfactant. A variety of small and medium-sized proteins were successfully recovered and detected using both electrospray and MALDI mass spectrometry over gel loads of 0.1 -10 μg. This technological tool is very important for extracting quality intact protein samples from polyacrylamide gels, from which accurate protein molecular weights and protein sequences can be obtained using intact molecules.

See more of this Session: Biomems and Microfluidics: Cell and Biomolecule Analysis
See more of this Group/Topical: Topical 3: 2009 Annual Meeting of the American Electrophoresis Society (AES)

# Probing Nanocolloids-Nanochannel Electrokinetic Interaction by Impedance Spectroscopy: A New Biosensing Platform

Gilad Yossifon, Department of Chemical and Biomolecular Engineering, University of Notre Dame, Notre Dame, IN

Peter Mushenheim, Department of Chemical and Biomolecular Engineering, University of Notre Dame, Notre Dame, IN

Hsueh - Chia Chang, Department of Chemical and Biomolecular Engineering, University of Notre Dame, Notre Dame, IN

We have designed and tested a combined nanochannel/nanocolloid biomolecule sensing platform based on a fundamental analysis of the underlying electrokinetic phenomena. The nanochannels are bounded by two micro-reservoirs to which the sample with dispersed nanocolloids are introduced. The nanocolloids are then driven electrophoretically, electro-osmotically or dielectrophoretically into the channel. Depending on the relative dimensions of the nanochannel cross-section, the nanocolloid size and the presence of hybridized molecules on the nanocolloids, the nanocolloids can accumulate at the nanochannel's entrance, pass through the nanochannel or even form a close-packed structure within the nanochannel [BMF 3 012001 (2009)]. Due to the small dimension of the channel, these events can significantly change the complex cross-channel impedance signal, as detected by the same microelectrodes that drove the nanocolloids into the nanochannels. Besides these non-linear characteristics associated with the nanocolloids, the nanochannels exhibit significant non-linear electrical responses of their own, such as vortex instability (PRL 101, 254501 (2008)), non-linear I-V curve (PRE 79, 046305 (2009)), voltage gating, channel cross-talk and current rectification effects. Understanding and being able to selectively filter and amplify the pertinent non-linear impedance signatures of this nanocolloid-nanochannel platform will be shown to provide sensitive biomolecular detection.

See more of this Session: Electrokinetic Behavior of Micro and Nanoparticles: Fundamentals and Applications
See more of this Group/Topical: Topical 3: 2009 Annual Meeting of the American Electrophoresis Society (AES)

# Experimental and Theoretical Comparison of Aspect-Ratio-Dependent Diffusion of CdSe Nanocrystals through Nanochannels

Louis J. Tribby, University of New Mexico, Albuquerque, NM

C. F. Ivory, Chemical Engineering, Washington State University, Pullman, WA

Frank Van Swol, Department of Chemical and Nuclear Engineering, University of New Mexico, Albuquerque, NM

**Sang M. Han**, University of New Mexico, Albuquerque, NM

The use of nanofluidic devices as a means of separating biomolecules, nanoparticles, and other small species is of fundamental importance. Experimentally obtained transport properties can be compared to theory and computer simulations to further elucidate our understanding of their transport properties at the nanoscale. In particular, we have evaluated toluene suspensions of CdSe nanocrystals (NCs) to determine the effect of their aspect ratio on diffusion constant and eventual equilibrium concentration in nanochannels (100 nm w x 400 nm d). To quantitatively monitor NC diffusion, we measure spectrally resolved photoluminescence (PL) from these NCs as a function of time and distance from the nanochannel inlet, using laser scanning confocal microscopy. The PL intensity is compared to a diffusion model that allows NC-wall adsorption. The comparison reveals that the hindered NC diffusion constant is approximately two orders of magnitude lower than that observed in microchannels and that the NC-wall adsorption is negligible. The NC equilibrium concentration, approximated by the NC concentration near the nanochannel inlet, also reveals strong aspect ratio dependence that compares well with a theory that accounts for physical interactions of NCs with channel geometry.

See more of this Session: Electrokinetic Behavior of Micro and Nanoparticles: Fundamentals and Applications
See more of this Group/Topical: Topical 3: 2009 Annual Meeting of the American Electrophoresis Society (AES)

## Electrophoretic transport of Nucleic acids through nano-structured surfaces

*Arnab Ghosh[1], Deepak Singh[2], Dr.R. Gurunath[2], Dr. Shantanu Bhattacharya[1,*]*
*(1) Department of Mechanical Engg., (2)Department of chemistry, Indian Institute of Technology-Kanpur*

Traditional electrophoresis methods like Gel electrophoresis, Zone electrophoresis , Capillary electrophoresis have the inability to separate nucleic acids with higher fragment sizes. A new alternative approach known as Surface electrophoresis was proposed and developed in the early part of this decade which used the electrostatic interaction between nucleic acid molecules and charged surfaces as a basis of performing fractionation. We have explored the behaviour of negatively charged molecules with nano-structured hydrophobic or hydrophilic surfaces or selective nature of nano-patterned and charged surfaces to such molecules which have seldom been explored. Molecular dynamic simulations used earlier to predict the behaviour of charged molecules translating across silicon or oxides have revealed some preliminary level explanation. We have been exploring nano-structured surfaces and their influence on nucleic acid mobility under electrophoretic fields. In this work we enumerate some of our findings on nanostructured porous silica (NPS) films. We perform translation of ds-DNA (2Kb and more) in variously patterned surfaces and observe the preferential selection of these surfaces as a function of porogen loading. The mobility studies are performed in a simple and novel flow cell specially developed in our laboratory for surface. The mobility of these stains is calculated by the help of a microscope, and a MATLAB code, which traces the boundary of the images of these stains and then calculate the centroid. The mobilities are further validated using molecular dynamic simulations.

## INTRODUCTION

DNA fractionation through size based electrophoresis have been regularly used for variety of diagnostics activities like gene sequencing, gene mapping, and clinical analysis/diagnosis so on so forth. Starting from the advent of the human genome project in early ninety's a great need was felt for rapid separation resulting in shorter analysis time. Gel electrophoresis [1], Zone electrophoresis [2,3], Capillary electrophoresis [4,5] (in long micron sized capillaries), and micro fabricated electrophoresis platforms for performing electrophoresis [6,7] were developed because of this great need for rapidity. However a major limitation of all this techniques has been the inability to separate at higher fragment sizes typically greater than 10 kilo base pairs. This is mostly owing to biased reptations which occur because of the entanglement of longer fragments of nucleic acids while traversing gel matrices. A new approach known as Surface electrophoresis was proposed and developed in the early part of this decade which used the electrostatic interaction between nucleic acid molecules and charged surfaces as a basis of performing fractionation. This approach gave an alternative to all the earlier developed separation and fractionation techniques for DNA molecules [8-11].

There have been several reports of surface electrophoresis using the chemical or topological properties of micro or nano-channels surfaces [8-11].

Whenever a DNA molecule is adsorbed on a flat surface there is loss of entropy due to the restrainment of DNA molecules on the surface, there is also an increase in energy due to adsorption phenomenon. The balance of these two energies promulgates the DNA segments to be present as either loops or trains. If you immerse the surface in a solution then the loops extend into the solution and the trains, which are contiguous segments adsorbed over the surface are held close to the surface. The shorter DNA molecules have less negative charge on them and so the repulsion force is not much .Therefore they form more number of loops extending into the solution then the longer DNA molecules which behave otherwise due to high columbic repulsion. When a planner electric field is applied the molecules adsorbed over the plane respond to this field as a function of their conformation states. As the longer DNA molecule has more number of train segments they experience greater resistance due to more prominent columbic attraction and thus move slowly. The shorter DNA molecule on the other hand has more loop structures freely extending into the solution hence they experience lower columbic attraction from the surface. Hence the shorter DNA molecules posses more mobility.

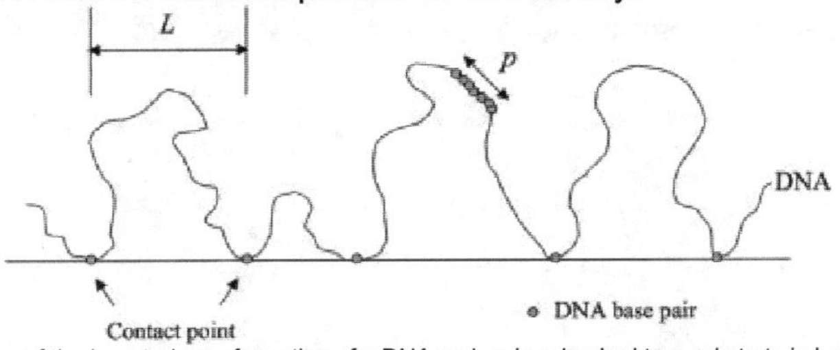


Figure 1: Schematic view of the loop-train conformation of a DNA molecule adsorbed to a substrate in buffered solution.

Although a very useful technique in recent times, not much analysis has been developed with this novel fractionating method. For example the behaviour of negatively charged molecules with nano-structured hydrophobic or hydrophilic surfaces or selective nature of nano-patterned and charged surfaces to such molecules has been seldom explored. Molecular dynamic simulations used earlier to predict the behaviour of charged molecules translating across silicon or oxides have revealed some preliminary level explanation. We have been exploring nano-structured surfaces and their influence on nucleic acid mobility under different electrophoretic fields. In this work we enumerate some of our findings on nanostructured porous silica (NPS) films. Also we have done some Molecular dynamic simulations to support our work.

## METHODS

***2.1 Substrate preparation (NPS):***

Nano porous materials have attracted much attention in recent years due to their great potential in a wide variety of applications such as photonics [12], catalysis [13] and semiconductor processing [14]. The most widely used methods to prepare nanoporous materials are the Sol-gel process which basically involves the gelation of a sol followed by controlled removal of the solvent to form the porous network, and the Porogen/template extraction methods which basically involve the removal of thermal labile species (Porogen) from the matrix by heating at higher temperatures.

We have used a simple technique for the formation of nanoporous films used by Rajagopalan et al. [15,16] which is a facile spin-on approach based on Polymethyl Silsesquioxane (PMSSQ) nanoparticles. PMSSQ nanoparticles were formed in the solution phase in the presence of a low molecular weight poly (propylene glycol (PPG). Nanoporous films were formed by spin coating this precursor solution onto silicon followed by a high temperature heating step. We can obtain films of different porosities from a single precursor solution by varying the processing conditions because the porosity of his films arises from interparticle separations. PPG acts as the pore generating agent during the final film formation. High temperature treatment step results in curing of the nanoparticles along with PPG removal for final film formation. High surface area of this material is attributed to the unique morphology of these films combining porosity with nanoparticulate nature. These films are hydrophobic with water contact angles in excess of 100 degrees thereby minimizing any moisture adsorption over extended periods of time. Figure 2 shows an SEM image of a NPS films cross section obtained with 50% PPG loading respectively. The pore sizes obtained in this film are in the range of 2-3 nm. It can be observed that the film is composed of nanoparticles with a relative uniform distribution.

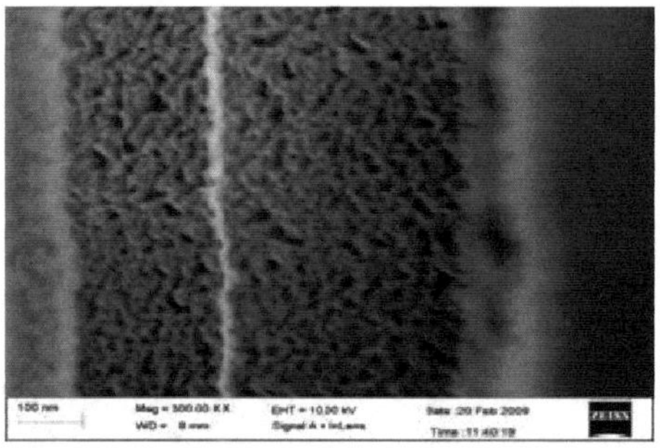

Fig 2: Crosse-sectional SEM image of 200nm NPS film prepared by spin coating after porogen removal by heating.

***2.2 Surface electrophoresis:***

The mobility studies are performed in a flow cell specially developed in our laboratory for surface electrophoresis. The experimental setup is shown in fig. 3(a). The cell is basically a small box made of PDMS where a DC voltage is applied to the metallic electrodes positioned at both flanks of a small holder platform. Two small pieces of PDMS are plasma bonded on the top of the substrate to create a narrow path to avoid diffusion of the DNA in all directions, and hence have a unidirectional movement in the direction of applied electric voltage. Fig. 3(b) shows the Si Substrate with PDMS pieces plasma bonded to it. The nano-structured film is coated on a silicon surface by using spin coating or dip coating protocols. The structuring of these films are done using high energy beams and also by heating them to high temperatures. A droplet of 2µl of DNA solution (2Kb and more) is deposited on the edge of the substrate closest to the negative electrode and allowed to air-dry for some time. The substrate is than placed between the two electrodes on the glass slide called as the Electrophoretic Strip. This is shown in Fig 3(c). The Electrophoretic strip is then inserted in the box through a groove which is filled with 1X Tris-acetate-EDTA (*TAE*) buffer at pH 8. Ethidium bromide dye was added to the buffer in order to stain the dsDna. The dye and buffer concentration was kept constant for all experiments described here. The electrodes are connected to a waveform generator which supplies the requisite signal to the flow-cell. Two different voltages (11 and 24 V) were applied at the electrodes at the ends of the cell, through the buffer solution. The images of the DNA blob were taken by a CCD camera mounted on top of a microscope as a function of time at a fixed position. The whole operation of imaging of the DNA blob is carried out within the field of view of a Nikon Trinocular fluorescence microscope and the stage is moved using a pre-calibrated micrometer lead-screw.

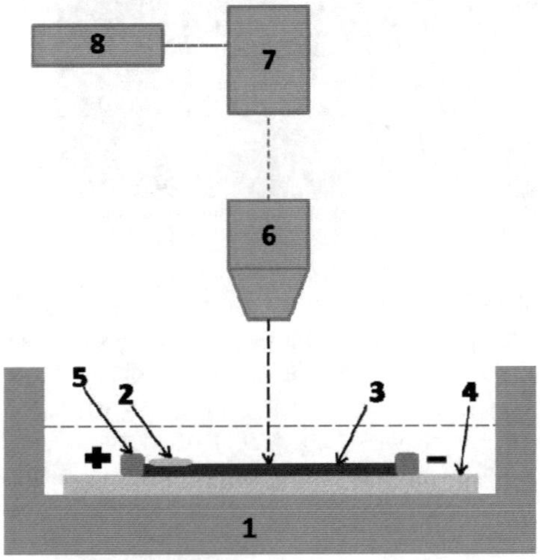


**Figure 3(a).** Surface electrophoresis experimental setup:

1. Sample cell,
2. DNA sample,
3. Substrate,
4. Glass slide,
5. Electrodes,
6. Fluorescence microscope,
7. CCD Camera,
8. Computer.

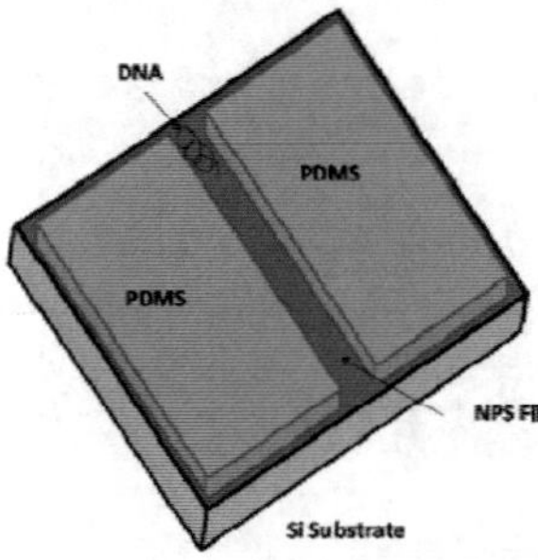


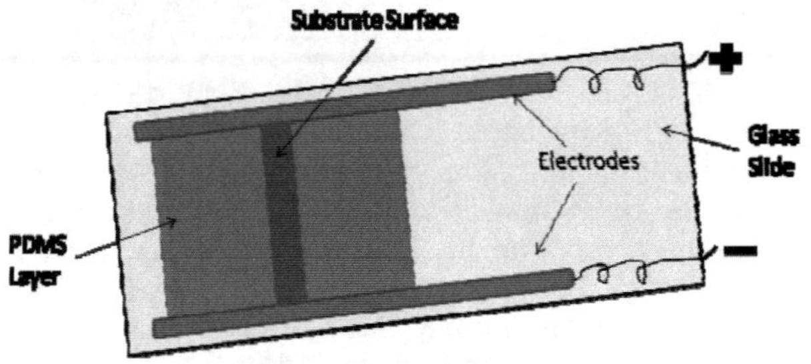


Figure 3(c) Electrophoretic Strip

Figure 3(b) Silicon Substrate with PDMS

## RESULTS AND DISCUSSION

All visualization is performed using a ETBR (Ethidium Bromide) intercalated ds-DNA physisorbed over the target. Fluorescence images of how this DNA looks over the electrophoretic strip is visualized by a Fluorescence microscope and shown in Fig. 4.

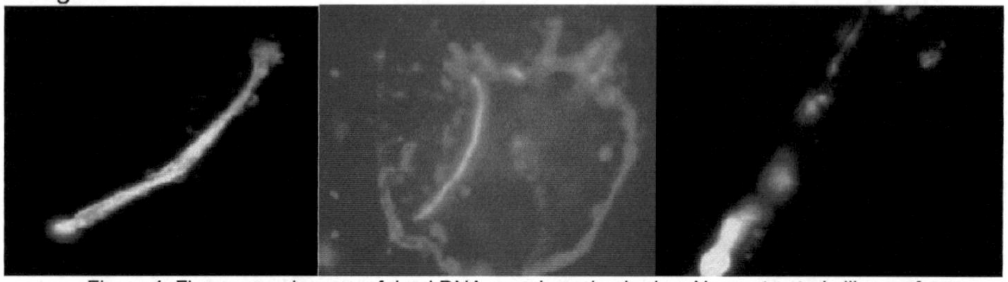

Figure 4: Fluroscence images of dyed DNA samples adsorbed on Nano-structed silica surfaces

We next apply the external electric field and study the motion of the intercalated DNA over various surfaces. Figure 5 shows 5 snapshots at 5 different time instants of the mobile DNA molecules over a silicon dioxide surface under the influence of an external electric field. As it is visible from these snapshots, clearly the DNA moves altogether as a blob and in the process is able to translate and rotate. To calculate the mobility of the DNA stains we have devloped a MATLAb code which traces the boundary of the DNA blob and then finds out the centroid of the blob. Hence we have fixed our microscope focus at a definate postion on top of the DNA stain and taken images at different intervals of time. The voltage applied between the electrodes is around 11V.

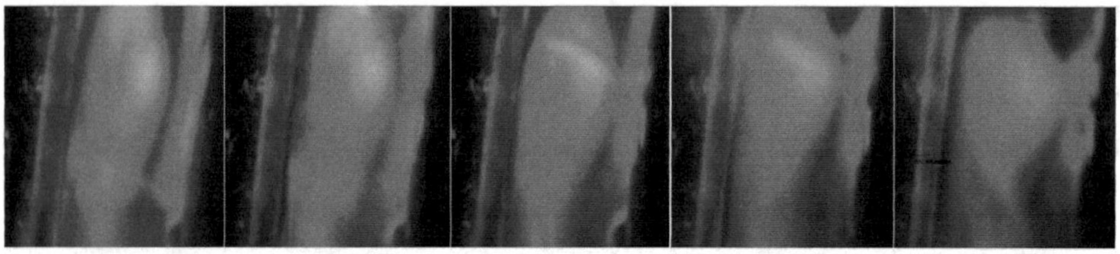



Figure 5: Fluroscence images of the dyed DNA samples adsorbed on silicon oxide surfaces at different time instants

After that we have used the MATLAB code to generate the boundary of the DNA blob and then calculate the centroid of the blob. This MATLAB code converts the image taken into binary format and then finds the start point of any boundary. Thereafter the code traces the whole boundary in terms of x and y position coordinates and calculates the centroid of the boundary as a unique (X,Y) coordinate. This enables us to find the movement of the centroid with time and subsequently calculate the mobility values. The velocity is determined diretly from the distance time data obtained by tracing the distance between the two poistions of the centroids at two time instants and diving this distance by the total time duration. This is then used to calculate the average mobilty of the DNA blob.

The voltage which we used here is 11 volts and the distance beteween the electrodes is 1 cm. The boundary that is traced by the MATLAB code is shown in the figure 6 below.

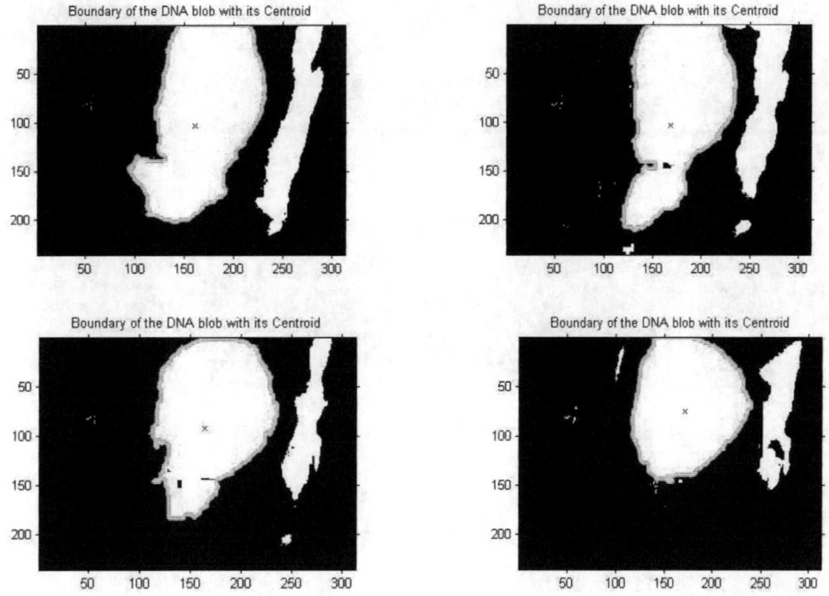


Figure 6: Boundary traced by MATLAB code of the Image of DNA stains

The mobility of the DNA blob is calculated as $4.29 \times 10^{-6}$ $cm^2/Vs$. It may be worth mentioning that the mobility of DNA in electrophoresis gels is in the range of $3.31 \times 10^{-5}$ $cm^2/Vs$. Our mobility on a surface is almost an order of magnitude lower than the gel mobility of the DNA stains. This may be attributed due to the fact that the DNA molecules are bound and constrained close to the surface which is not the case in gel electrophoresis where the only interaction that they share is with the gel walls. The movement of the centroid with time is shown in the graph below.

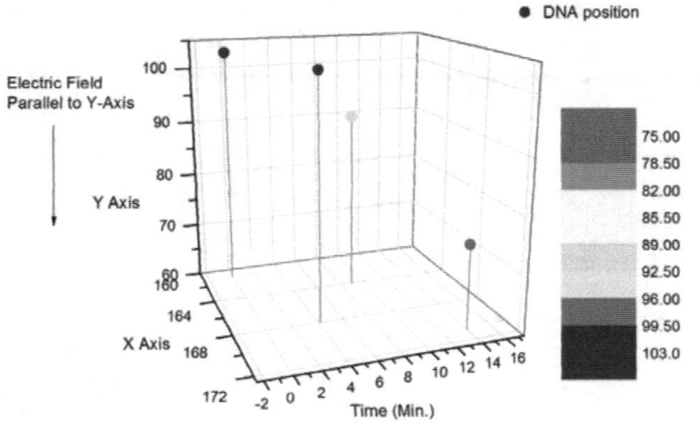


Figure 7: Graph showing the displacement of centroid of a 2kb dsDNA stain with time

We next use a 1 Kb DNA ladder solution to study its motion over silicon dioxide surface. The dc voltage applied in this case is 24V. Figure 8 shows 7 snapshots at 7 different time instants of the mobile DNA molecules over a silicon dioxide surface under the influence of an external electric field.

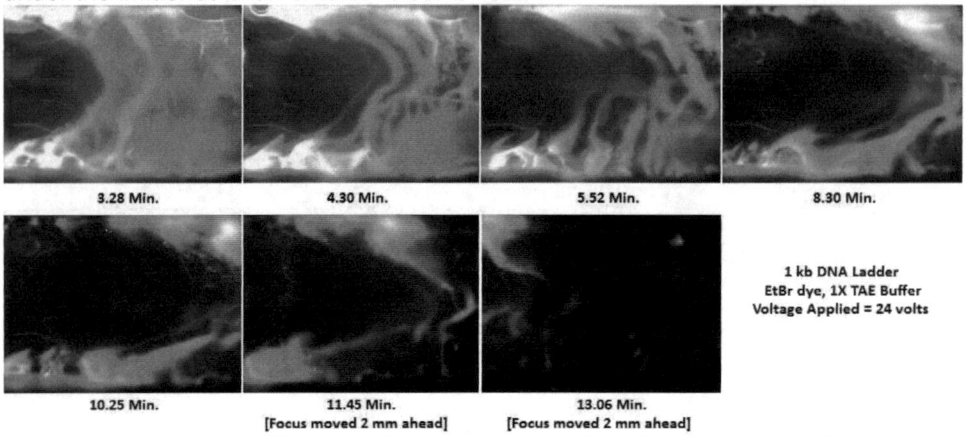


Figure 8: Fluroscence images of the dyed DNA samples adsorbed on silicon oxide surfaces at different time instants

After that we have used the MATLAB code to generate the boundary of the DNA blob and then calculate the centroid of the blob. The voltage which we used here is 24 volts and the distance beteween the electrodes is 1 cm. The boundary that is traced by the MATLAB code is shown in the figure 9 below.

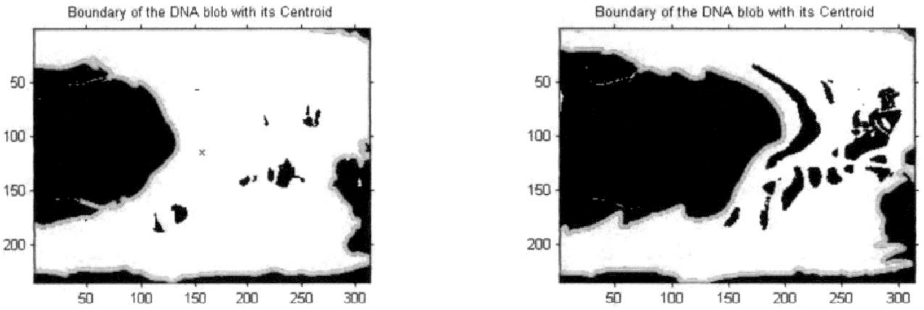

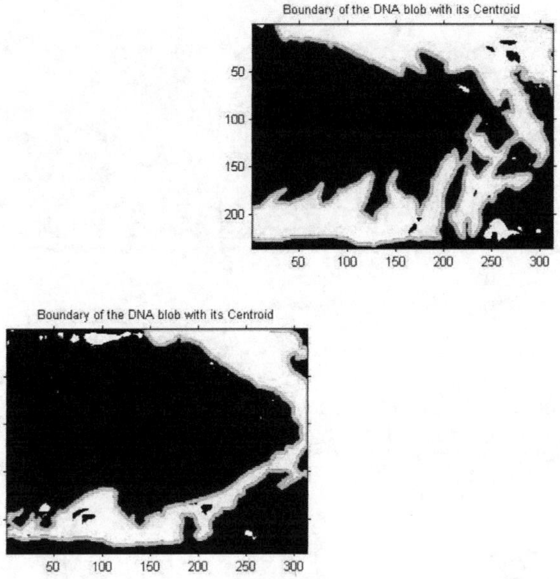


Figure 9: Boundary traced by MATLAB code of the Image of 1 Kb DNA ladder stains

The mobility of the DNA blob is calculated as 3.96x $10^{-6}$ $cm^2/Vs$. The mobility obtained of 1kb ladder is less than that of the previous case. This may be attributed due to the fact that as the size of the DNA molecules is more and thus the interaction with the surface is more resulting in overall slower movement of the DNA blob. The movement of the centroid with time is shown in the graph below.

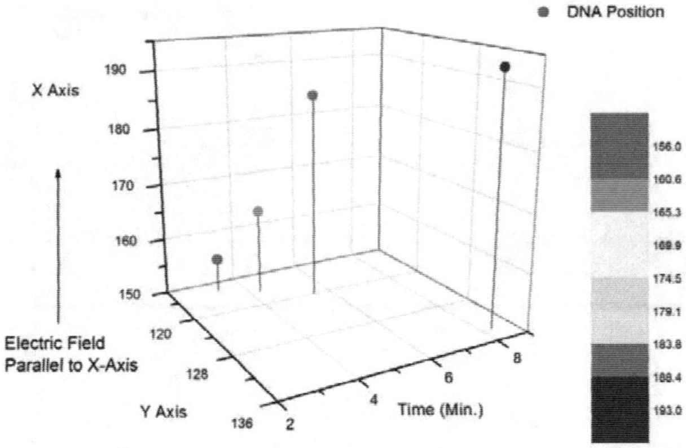


Figure 10: Graph showing the displacement of centroid with time

Next we explored the nucleic acid mobility under electrophoretic fields on Nano Porous Silica surfaces. We use a 1 Kb DNA ladder solution to study its motion over Nano Porous Silica surface. The dc voltage applied in this case is 11V and the distance beteween the electrodes is 1 cm. Figure 11 shows 5 snapshots at 5 different time instants of the mobile DNA molecules over a Nano Porous Silica surface under the influence of an external electric field.

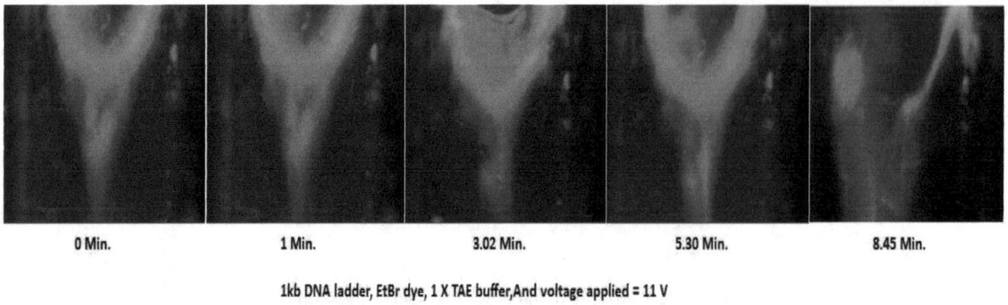


Figure 11: Fluroscence images of the dyed DNA samples adsorbed on Nano Porous Silica surfaces at different time instants

After that we have used the MATLAB code to generate the boundary of the DNA blob and then calculate the centroid of the blob. The boundary that is traced by the MATLAB code is shown in the figure 12 below.

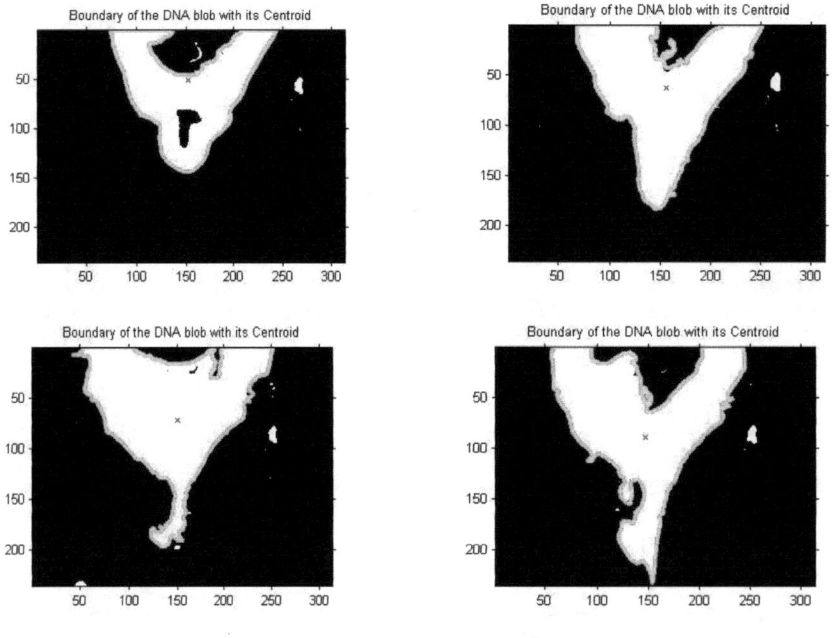


F

igure 12: Boundary traced by MATLAB code of the Image of 1 Kb DNA ladder stains

The mobility of the DNA blob is calculated as 6.01 x $10^{-6}$ $cm^2/Vs$. The mobility obtained of 1kb ladder on Nano Porous Silica surface is more than that on the silicon Oxide surface. This may be due to the fact that the interaction between Nano Porous Silica surface and the DNA molecules is more as compared to the interaction between silicon dioxide surface and DNA molecules, resulting in overall faster movement of the DNA blob on Nano Porous Silica surfaces. The movement of the centroid with time is shown in the graph below.

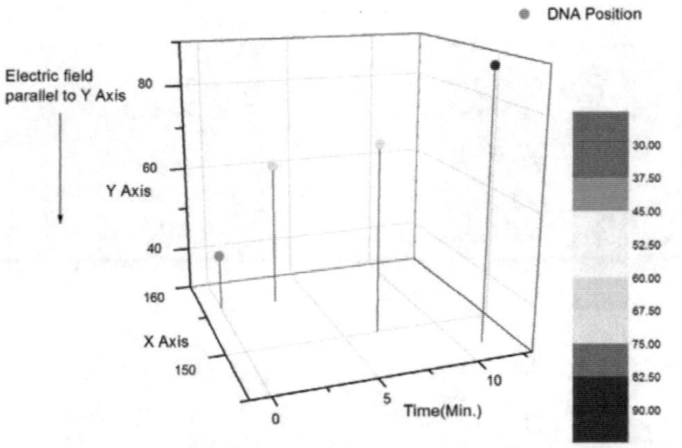


Figure 13: Graph showing the displacement of centroid with time

Next we see what is the effect of increasing the electric field on mobility of the DNA molecules. Here we have applied a dc voltage of 24 Volts. The DNA sample is the same as used the previous case i.e. 1 kb DNA ladder. Figure 14 shows 5 snapshots of the DNA stains at different time instants.

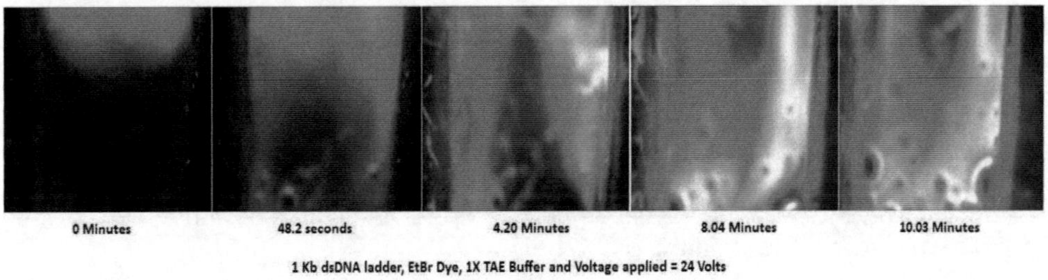


Figure 14: Fluroscence images of the dyed DNA samples adsorbed on Nano Porous Silica surfaces at different time instants

After that we have used the MATLAB code to trace the boundary of the DNA blob and then calculate the centroid of the blob. The boundary that is traced by the MATLAB code is shown in the figure 15 below.

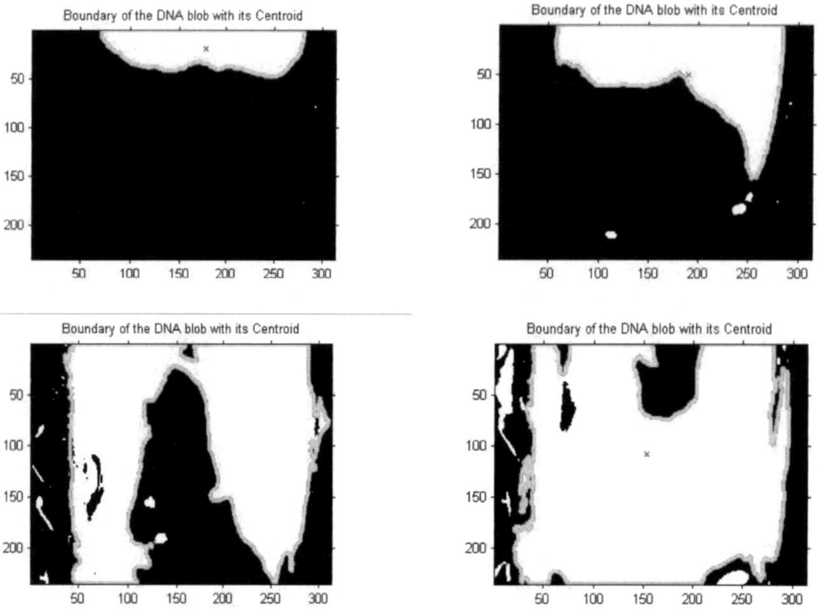


Figure 15: Boundary traced by MATLAB code of the Image of 1 Kb DNA ladder stains

The mobility of the DNA blob is calculated as 9.958 x $10^{-6}$ $cm^2/Vs$. The mobility obtained of 1kb ladder on Nano Porous Silica surface with 24 V is more than the mobility obtained due to 11 V. This is quite apparent because as the voltage is increased the force acting on the DNA molecules increases which causes it to move faster, resulting in overall faster movement of the DNA blob. The movement of the centroid with time is shown in the graph below.

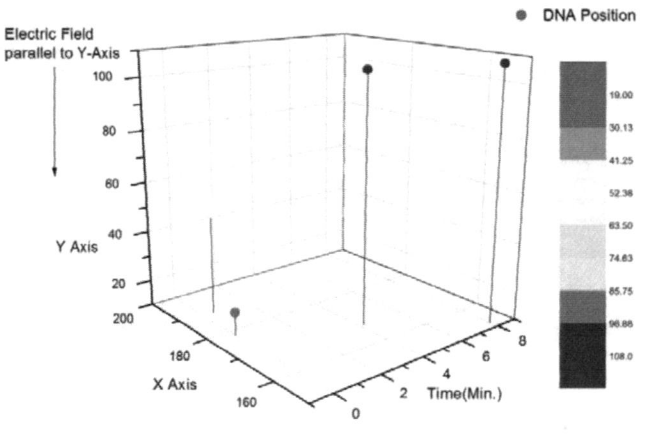


Figure 16: Graph showing the displacement of centroid with time

Molecular dynamics simulations indicate that the DNA molecules (represented by bead string models) take a finite time to reach thermodynamic equilibrium with a surface the final step of which is a loop train model. [ Fig. 17]

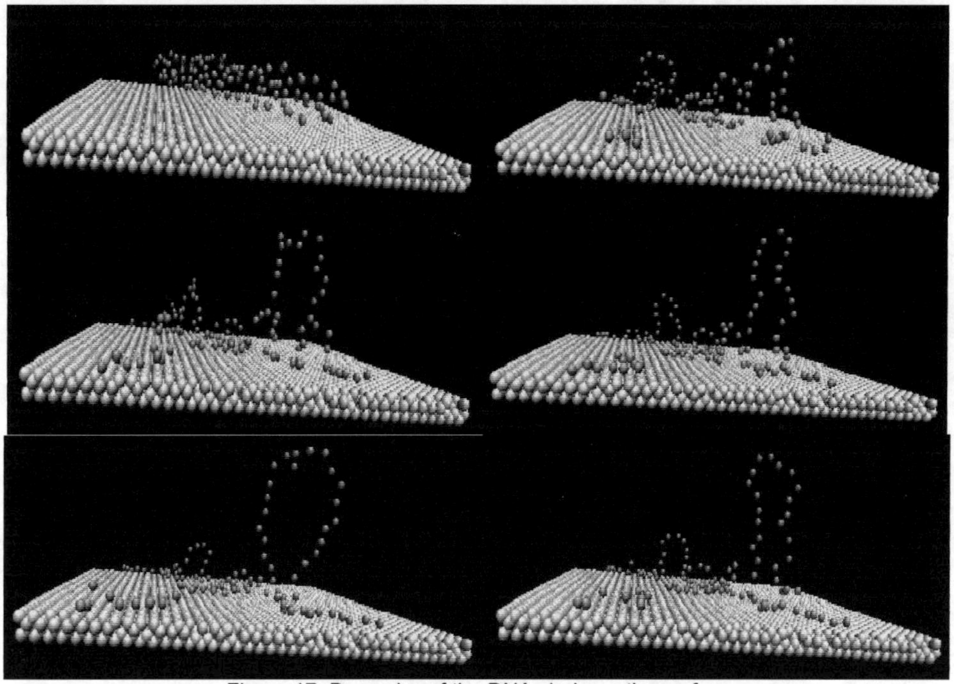

Figure 17: Dynamics of the DNA chain on the surface

This is primarily due to vanderwalls forces, loss of entropy due to the restrainment of DNA molecules on the surface, and also an increase in energy due to adsorption on the surface. As we put electric field in this model we do see a translatory motion of the molecules towards the positive polarity. However, this model can only predict a single molecule translatory behavior. As in actual experiments we cannot simulate the movement of DNA clumps as the computational requirements of such a study will be enormous. On a single moelcule basis the movement of DNA can be peridictive to a reasonable extent of parameters like mobility and we are trying to compute these using MD simulations.

The mobility data is calculated from the velocity and electric field by the one dimensional mobility equation. Further mobilities are investigated over plain and nanostructured surfaces. From fig. it is clear that the mobility on the Nanoporous silica surface is more as compared to normal silicon dioxide layer. Further between the two different surfaces the mobility enhancement is found on nanoporous silica surface. At this time we hypothesize the nanoporous silica surface to be charged and we are further trying to find the zeta potential of a nanoporous silica surface in contact with a pH of 8.3 which is the pH of 1XTAE buffer. Due to the surface negative charge there is a tendency of repelling the negatively charged DNA and thus makes the molecules more mobile. We are also trying to understand this phenomenon using the MD simulations on a single molecule basis.

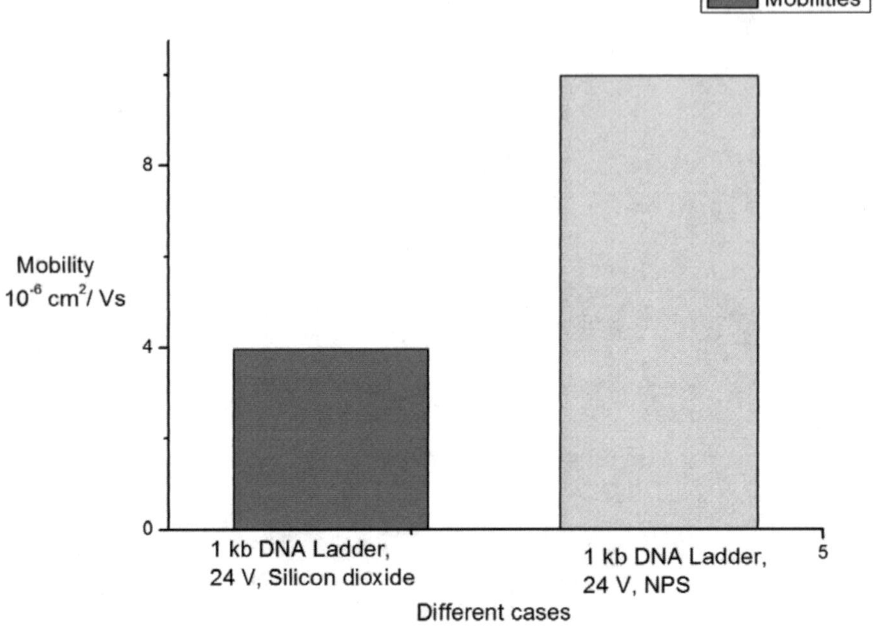


Figure 18: Mobility calculations for DNA molecules over different surfaces under different operating voltages

## CONCLUSIONS

We have developed a unique protocol to monitor and measure surface electrophoresis of charged DNA molecules on a surface. Further comparison between nanostructured silica surface and an ordinary silicon dioxide surface reveals that mobility increases between the two by almost 3-4 times. The real reasons are explored as yet but it can be hypothesized that the nanoporous silica surface develops a negative surface charge owing to the proximity of the 8.4 pH 1XTAE medium in which the film is immersed after absorption on DNA molecules on its surface. Further MD simulations are being carried out on a single molecule basis and we have observed that the molecules need a finite amount of time to equilibrate over the surface in question and then start moving after application of an external electric field with different mobilities.

Further studies are on the explore the real reasons of the enhanced mobility in the nanostructured silica surface.

## References:


1) Z. Djouadi, S. Bottani, M. A. Duval, R. Siebert, H. Tricoire, L. Valentin, "Dynamics of single-stranded DNA migration in denaturing polyacrylamide slab- gel electrophoresis", Vol. 22(16), pp.3527-3532, Electrophoresis(2001).
2) M. G. Janini, R. J. Fisher, L. E. Henderson, H. J. Issaq, "Application of capillary zone electrophoresis for the analysis of proteins, protein-small molecules, and protein-DNA interactions", Vol.18, pp. 3617-3628, Journal of Liquid Chromatography (1995).
3) M.Chiari, M. Nesi, P. G. Righetti, "Capillary zone electrophoresis of DNA fragments in a novel polymer network: poly (N-acryloylaminoethoxyethanol)", vol.15 (5), pp. 616-622, Electrophoresis (1994).
4) C. Patrick, "Capillary electrophoresis, Theory and Practice", CRC press, London, pp. 200-207, 1993.
5) S Bhattacharya, "Economy of Advance Trainings: Basics, Concepts, Methods", VDM Verlag Dr. Muller Aktiengesellschaft & Co. KG.
6) E. T. Lagally, H. T. Soh, " Integrated genetic analysis of Microsystems", Critical reviews in solid state and materials science, Vol. 30, pp. 207-233 , 2005.
7) V. M. Umag, R. D. Elms, R. C. Lo, "Micro-fabricated electrophoresis systems for DNA sequencing and genotyping applications: current Technology and future directions, Phill. Trans. R. Soc. Lond. A, Vol. 362, pp. 1105-1129, 2004.
8) N. Pernodet, V. Samuilov, K. Shin, J. Sokolov, M. H. Rafailovich, D. Gersappe, and B. Chu: Phys. Rev. Lett. 85, 5651 (2000)
9) Y. S. Seo, V. A. Samuilov, J. Sokolov, M. Rafailovich, B. Tinland, J. Kim, and B. Chu: Electrophoresis 23, 2618 (2002)
10) Y.-S. Seo, H. Luo, V. A. Samuilov, M. H. Rafailovich, J. Sokolov, D. Gersappe, and B. Chu: Nano Lett. 4 , 659 (2004).
11) Y. Kuo and H. H. Lee: Electrochem. Solid-State Lett. 4 , H23 (2001).
12) K. Biswas, S. Gangopadhay, H. C. Kim, R. D. Miller, Thin Solid Films 2006, 514,350
13) Y. Y. Jackie, AIChE journal 2000, 46, 1962
14) M. Ree, J. Yoon, K. Heo, Journal of Materials Chemistry, 2006,16, 685
15) A.A. Lubguban, J.A. Lubguban, M.T. Othman, R.V. Shende, S. Gangopadhyay, R.D. Miller, W. Volksen, H.-C. Kim, Thin Solid Films, 2008. 516, 4733
16) T. Rajagopalan, B. Lahlouh, J. A. Lubguban, N. Biswas, and S. Gangopadhyay, Applied Physics Letters, 2003, 82, 24

# Dielectrophoresis of Colloidal Suspensions of SWNTs: Formation of 1-D Assemblies

**Mainak Majumder**, Dept. of Chemical and Biomolecular Engineering, Rice University, Houston, TX

Budhadipta Dan, Dept. of Physics and Astronomy, Rice University, Houston, TX

Michelle Li, Dept. of Chemical and Biomolecular Engineering, Rice University, Houston, TX

Howard K. Schmidt, Dept. of Chemical and Biomolecular Engineering, Rice University, Houston, TX

Matteo Pasquali, Dept. of Chemical and Biomolecular Engineering, Rice University, Houston, TX

SWNTs are nanoscale materials with extreme anisotropy and electrical polarizability, whereas dielectrophoresis is the transport of polarizable particles in an electrical field gradient. With a view to fabricate 1D linear assembly from SWNT colloidal suspensions stabilized by surfactants such as Sodium dodecylbenzene sulphonate and Sodium deoxycholate, we have studied the dielectrophoretic behavior in a plate and pin geometry using in-situ optical visualization. Depending upon the nature of the surfactant, operating voltage and frequency, 1D assemblies ranging in lengths from a few microns to several hundred microns have been successfully fabricated. These assemblies have been characterized by optical microscopy, scanning electron microscopy, and current-voltage measurements.

See more of this Session: Electrokinetic Behavior of Micro and Nanoparticles: Fundamentals and Applications
See more of this Group/Topical: Topical 3: 2009 Annual Meeting of the American Electrophoresis Society (AES)

# Corrosion and Sulfate Attack Mitigation in Concrete Using Electrokinetic Nanoparticle Treatment

Henry E. Cardenas, Department of Mechanical Engineering, Louisiana Tech University, Ruston, LA

Kunal V. Kupwade-Patil, Department of Mechanical Engineering, Louisiana Tech University, Ruston, LA

Jinko Kanno, Department of Mathematics and Statistics, Louisiana Tech University, Ruston, LA

James Phillips, Department of Mathematics and Statistics, Louisiana Tech University, Ruston, LA

**Daniela Mainardi**, Department of Chemical Engineering, Louisiana Tech University, Ruston, LA

This work demonstrates a new method to rehabilitate damaged concrete using electrokinetic insertion of pozzolanic nanoparticles. The intent was to use the electric field to simultaneously extract deleterious species while inserting nanoparticles as a means of inhibiting the return migration of these agents. Reinforced concrete deteriorates due to the chloride induced corrosion of the steel reinforcement. The hardened cement binder phase of concrete is susceptible to sulfate attack. In this study reinforced concrete specimens were exposed to sulfates and others to chlorides. The chlorides and sulfates were extracted concurrent with the initiation of electrokinetic nanoparticle (EN) treatment. EN treated specimens exhibited 30% decreases in porosity and 33 % increases in strength in the sulfate exposed specimens. Among the chloride exposed specimens EN treatment yielded significantly lower corrosion rates and 50-75 % reductions in porosity. It was observed that the level of porosity reduction appeared to correlate to the sizes of species transported. Analytical modeling and a computer simulation were used to understand the nature of porosity reductions achieved via electrokinetic nanoparticle transport in these systems.

See more of this Session: Electrokinetic Behavior of Micro and Nanoparticles: Fundamentals and Applications
See more of this Group/Topical: Topical 3: 2009 Annual Meeting of the American Electrophoresis Society (AES)

# Pressure Measurement in Microsystems Using Volume-Displacement of Nanoparticle Suspension

**Hyewon Lee**, Chemical & Biomolecular Engineering, Georgia Institute of Technology, Atlanta, GA

Kwanghum Chung, Chemical & Biomolecular Engineering, Georgia Institute of Technology, Atlanta, GA

Hang Lu, Chemical & Biomolecular Engineering, Georgia Institute of Technology, Atlanta, GA

Pressure is widely used to control fluid motions in microsystems. To achieve the accurate pressure measurement inside of microchannels, a number of detection methods have been developed. While most of these methods show high degrees of accuracy, they are not widely applicable for many microsystems due to a few drawbacks, such as size constraint, complications in fabrication, and expensive off-chip equipment. To overcome these limits, we present here a simple image-based method to measure pressure using volume displacement of fluorescent particle suspension. We fabricated microdevices consisting of two layers and polydimethylsiloxane (PDMS) membrane in between: bottom layer includes a flow channel and top layer contains a detection channel filled with fluorescent particles. Pressure increase in the flow channel deflects the PDMS membrane and its deformation can be quantified by software developed in-house. The area in which particles are in- and out-of-focus is measured as a function of pressure. This simple image processing is easy to use and produces data easy to interpret. Using this method, we accurately measured pressure in the range of 0.1 ~ 10 psi. This sensing range can be extended by tuning focal planes or reference pressures. Furthermore, because pressures in various parts of the chip can be transduced and the signals transferred to a single location, we can simultaneously detect pressures in multiple locations of microsystems. This multiplex pressure measurement can improve the throughput of microfluidic devices especially for highly integrated multi-functional devices. Compared to existing techniques, our pressure sensor is very easy to integrate without complicated fabrication processes and its performance can be optimized by simply changing geometry and thickness of the membrane.

See more of this Session: Electrokinetic Behavior of Micro and Nanoparticles: Fundamentals and Applications
See more of this Group/Topical: Topical 3: 2009 Annual Meeting of the American Electrophoresis Society (AES)

## INSULATOR-BASED DIELECTROPHORESIS WITH DIRECT CURRENT ELECTRIC FIELDS

**Javier L. Baylon-Cardiel,[1] Ana V. Chávez-Santoscoy,[1] Sergio O. Martínez-Chapa,[1] Claudia Reyes-Betanzo,[2] and Blanca H. Lapizco-Encinas[1,*]**

*[1]Tecnológico de Monterrey, Monterrey, NL, México. [2]Instituto Nacional de Astrofísica, Óptica y Electrónica, Santa María Tonanzintla, Puebla, México*

### Abstract

A mathematical model is used to characterize the performance of a device for insulating structures dielectrophoresis. Using a simple relation involving the electric field and other parameters, trapping by dielectrophoresis is estimated. Results from simulation and from experiment are compared.

### Introduction

Dielectrophoresis (DEP) is a non-destructive electrokinetic technique usually employed in micro-scale devices for the manipulation of particles, by means of the polarization effects due to the presence of a non-uniform electric field. DEP can be obtained with the application of alternating current (AC) and direct current (DC) electric fields.[1-3] This technique has been successfully implemented for the separation and concentration of a wide variety of particles, ranging from macromolecules to parasites.[1,4,5] Traditionally, DEP has been used employing a combination of arrays of metallic electrodes and AC electric fields. The electrode-based approach for DEP allows to obtain high electric field gradients with low applied electric potentials.[5-7] However, significant electrolysis can be produced, and massive parallel systems aimed to generate high throughput are prohibitive, due to the high cost of the electrode fabrication.[8]

Cummings and Singh proposed a different approach for DEP, by employing an array of cylindrical posts embedded in a microfluidic channel.[9] This approach is referred to as insulator-based DEP (iDEP). Here, the electric potential is applied across an array of insulating structures, employing only two electrodes that straddle the array of insulating structures. When an electric potential is applied, the electric field lines are bent in the space between the insulating structures, generating a non-uniform electric field required for DEP.[9] When iDEP implemented employing DC electric fields in substrates with surface charge (*e.g.* glass), other electrokinetic transport mechanisms are also present: electrophoresis and electroosmotic flow. In order to concentrate particles, DEP has to overcome the electrokinetics (electrophoresis and electroosmosis).[9] In this work, the performance of an iDEP device was modeled using the computational software *COMSOL Multiphysics*. The device modeled consisted of a microfluidic channel containing an array of cylindrical insulating structures. 1-um-diameter polystyrene microspheres were selected as probe particles. The purpose of the model was the prediction of the location and magnitude of zones where trapping of the particles due to DEP occur. The trapping regions were studied as a function of the properties of the suspending medium. Furthermore, simulation results were confirmed by experiments.

## Theory

The electrical potential applied along the microfluidic channel, can be obtained from the solution of the Laplace equation:

$\nabla^2 V = 0$ (1)

with boundary conditions:

$\vec{n}\cdot\vec{J} = 0$ at the boundaries and in the posts (2)

$V = V_{in}$ at inlet (3)

$V = 0$ at oulet (4)

where $\vec{n}$ in the normal vector to the surface, $\vec{J}$ is the electrical current and $V_{in}$ is the electrical potential applied between the posts.

The contribution of the applied electric field to the particle motion can be described by the electrokinetic (EK) velocity. The EK velocity is related to electric field as:

$\vec{u}_{EK} = \mu_{EK}\vec{E}$ (5)

where $\mu_{EK}$ is the electrokinetic mobility, and it takes account for the contributions electroosmosis and electrophoresis to the particle transport. Moreover, the motion due to the DEP force exerted on particles can be described by the DEP velocity:

$\vec{u}_{DEP} = -\mu_{DEP}\nabla E^2$ (6)

where $\mu_{DEP}$ is the dielectrophoretic mobility.

In order to obtain trapping by DEP, its contribution to the particle transport must overcome diffusion and EK contributions. When the net flux of particles along electric field lines, a condition for DEP trapping can be derived, and it is a relation expressed as

$\dfrac{\mu_{DEP}\nabla E^2}{\mu_{EK}E^2}\cdot\vec{E} > 1$ (7)

## Materials and Methods

### Mathematical model

The geometry of a microfluidic channel was captured into *COMSOL Multiphysics* software. The micro channel considered was 10.12-mm long, and 2-mm wide, containing an array of 32 insulating posts, distributed in 4 rows of 8 posts each. The posts had a diameter of 470 μm, arranged 510 μm center-to-center (Figure 1). Particles with a diameter of 1 μm, and conductivity of 10 $\mu Scm^{-1}$ were considered. Also no surface charge was taken into account in the model. A triangular mesh was built in order to solve the Laplace equation using the *Lagrange element* implemented in the software. From the obtained numerical solution it is possible to obtain values for the electrical field and the gradient of $E^2$.

### Experimental

The microfluidic channel employed for experimentation was fabricated from Schott 263 glass wafers. A top wafer containing drilled reservoirs and a bottom wafer contain wet-etched microchannels. The geometry of the microchannel employed is described above and shown in Figure 1. The particles employed were carboxylated-modified fluorescent green-yellow 1-μm-diameter microspheres (Invitrogen, Carlsbad, CA, USA), with a concentration of 7.639 x $10^{10}$

microspheresml$^{-1}$. These particles contained a fluorescent dye with excitation/emission wavelengths of 505/515 nm. Image and video capturing for was done employing an inverted microfluidic microscope SVM340 with an integrated video camera (LabSmith, Livermore, CA, USA). A high voltage power supply HVS448 (LabSmith, Livermore, CA, USA) was also used to apply DC electric fields.

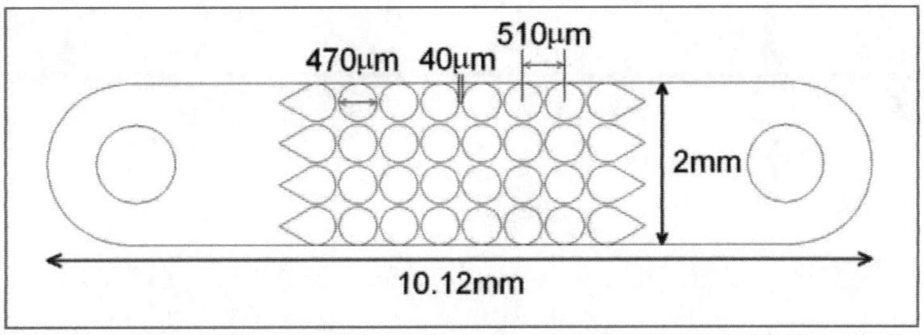


**Figure. 1.** Schematic representation of the microchannel employed for the mathematical model and experimentation.

### Results and discussion

#### Mathematical modeling

Figure 2 presents the distribution of the electric field and the gradient of $E^2$ between the array of insulating posts, when an electric potential of 500 V is applied. Regions of higher and lower electric field intensity were generated. It can be seen in Figure 2a that the greater magnitude of the electric field is located at the narrow region between the posts. Figure 2b shows that the electric field gradient of $E^2$ has the greatest magnitude at the narrow region between the cylindrical posts, close the post borders. It can be seen how regions of higher and lower magnitude relate to the distribution of electric field between insulating posts. Regions of higher magnitude of the gradient of $E^2$ are generated when the insulating posts are close to each other in the horizontal direction.

#### Dielectrophoretic Trapping

Figure 3 shows how trapping regions obtained with the computer model relate to the trapping regions observed in experiments. Figure 3a shows the regions of trapping for 1-μm particles when a field of 750 V is applied and a medium with pH of 6 and conductivity of 100 μScm$^{-1}$ is employed. Figure 3b shows the experimental results obtained under the same conditions. Figures 3c and 3d show the trapping predicted by the model and what is obtained in the experiment, for a pH of 6 and 50 μScm$^{-1}$. As the conductivity of the suspending medium increases, the trapping region becomes greater. For pH 9 and conductivity 100 μScm$^{-1}$, the DEP trapping region decreases, as shown in Figures 3e and 3f.

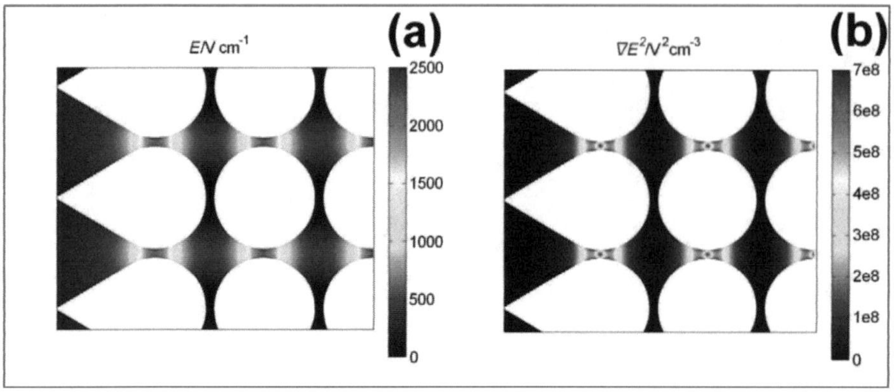


**Figure 2**. Simulation results. (a) Distribution of the electric field magnitude (b) Distribution of the magnitude of the gradient of $E^2$

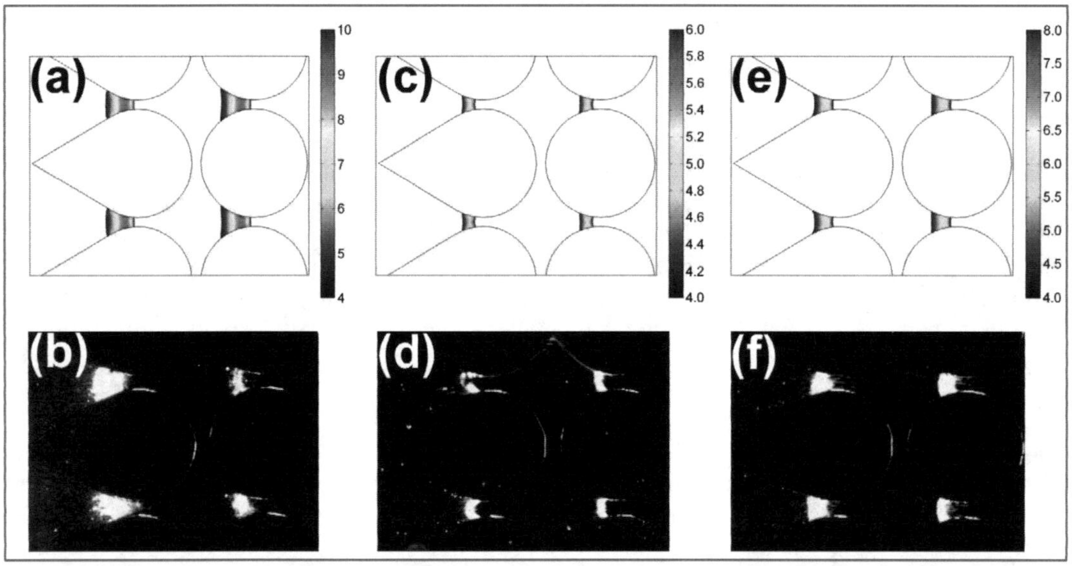


**Figure 3**. Predicted trapping zones with simulation and experimental observations, with and applied electric potential of 750 V and under different suspending mediums. (a) simulation with pH 6, 100 $\mu Scm^{-1}$ (b) experimental result (c) simulation with pH 6, 50 $\mu Scm^{-1}$ (d) experimental result (e) simulation with pH 9, 100 $\mu Scm^{-1}$ (f) experimental result.

## Conclusions

By employing a relation involving EK and DEP velocities, the regions of DEP trapping along an array of insulating cylindrical structures can be predicted. The results obtained with the computational simulation relate accurately with the experimental observations. Furthermore, the model was tested by varying the properties of the suspending medium, and it was shown that the model can account for these changes.

## Acknowledgments

The authors wish to thank the financial support given by the grant CONACYT-CB-2006-53603, and *Cátedra de Investigación* of *Tecnológico de Monterrey* (CAT 142).

## References

1 Voldman J., (2006), "Electrical forces for microscale cell manipulation," *Ann. Rev. Biomed. Eng.* 8, pp.425-454.

2 Gonzalez C.F., Remcho V.T., (2005), "Harnessing dielectric forces for separation of cells, fine particles and macromolecules.," *J. Chromatogr. A* 1079(1-2), pp.59-68.

3 Jones T.B., (1995), "Electromechanics of Particles," Cambridge University Press, USA.

4 Lapizco-Encinas B.H., Rito-Palomares M., (2007), "Dielectrophoresis for the manipulation of nanobioparticles," *Electrophoresis* 28(24), pp.4521-4538.

5 Gascoyne P.R.C., Vykoukal J., (2002), "Particle separation by dielectrophoresis," *Electrophoresis* 23(13), pp.1973-1983.

6 Ermolina I., Morgan H., (2005), "The electrokinetic properties of latex particles: comparison of electrophoresis and dielectrophoresis," *J. Colloid Interface Sci.* 285, pp.419–428.

7 Hughes M.P., (2002), "Strategies for dielectrophoretic separation in laboratory-on-a-chip systems," *Electrophoresis* 23, pp.2569-25582.

8 Simmons B.A., McGraw G.J., Davalos R.V., Fiechtner G.J., Fintschenko Y., Cummings E.B., (2006), "The Development of Polymeric Devices as Dielectrophoretic Separators and Concentrators," *MRS Bulletin* 31, pp.120-124.

9 Cummings E.B., Singh A.K., (2003), "Dielectrophoresis in Microchips Containing Arrays of Insulating Posts: Theoretical and Experimental Results," *Anal. Chem.* 75(18), pp.4724-4731.

# Negative and Positive Dielectrophoresis of Particles in Curved Microchannels

**Juan Nieto**, Mechanical Engineering, Clemson University, Clemson, SC

Gyunay Keten, Mechanical Engineering, Clemson University, Clemson, SC

Erl Ibarra, Mechanical Engineering, Clemson University, Clemson, SC

Christopher Church, Mechanical Engineering, Clemson University, Clemson, SC

Xiangchun Xuan, Mechanical Engineering, Clemson University, Clemson, SC

We study particle electrophoresis and dielectrophoresis in curved microchannels. It is observed that 1 um particles are focused to either the centerline or the sidewalls of the channel depending on the concentration of the suspending medium. This phenomenon is a consequence of the negative or positive dielectrophoresis of particles induced by the non-uniform electric field intrinsic to curved channels. We utilize such distinct focusing trends to separate particles in curved microchannels by size.

See more of this Session: Electrokinetic Behavior of Micro and Nanoparticles: Fundamentals and Applications
See more of this Group/Topical: Topical 3: 2009 Annual Meeting of the American Electrophoresis Society (AES)

# Thermal Stabilization of Tissue Samples Using the Denator Stabilizor T1: The Inactivation of Proteases and Elimination of Proteolytic Fragments On Two-Dimensional Gels

Gary B. Smejkal, Laboratory for Innovative Translational Technologies, Harvard Clinical and Translational Science Center, Boston, MA

Alexander J. Trachtenberg, Laboratory for Innovative Translational Technologies, Harvard Clinical and Translational Science Center, Boston, MA

J. Robert Chang, Laboratory for Innovative Translational Technologies, Harvard Clinical and Translational Science Center, Boston, MA

John Lindsay, Denator, Biotech Center, Arvid Wallgrens Backe 20, Gothenburg, Sweden

Mats Borén, Denator, Biotech Center, Arvid Wallgrens Backe 20, Gothenburg, Sweden

Soulafa A. Almazrooa, Laboratory for Innovative Translational Technologies, Harvard Clinical and Translational Science Center, Boston, MA

Winston P. Kou, Laboratory for Innovative Translational Technologies, Harvard Clinical and Translational Science Center, Boston, MA

An understanding of cellular processes is facilitated only when methods such as two-dimensional gel electrophoresis (2DGE) are capable of isolating without bias the entire protein constituency of cells such that the analysis can provide a comprehensive representation of the proteome. The regulated expression of proteins as it relates to normal or disease processes, the efficacy of therapeutic regime, or the identification of phenotypes requires that specific proteins are accurately quantified. Therefore, reliable methods of sample preparation become necessary to produce meaningful results. Since proteases regulate many cellular processes including the clearance of down-regulated proteins and the cleavage of precursors to produce functional proteins, it is essential to completely abolish all enzymatic activity to elicit a realistic "snapshot" of the proteome as it occurs in vivo, both qualitatively and quantitatively. Further, the phosphorylation states of proteins may be altered if enzymes involved in phosphorylation and dephosphorylation are not promptly inactivated. Typically, proteolytic activity is curtailed by rapidly freezing the sample and by the inclusion of protease inhibitors and metal chelators. However, freezing preserves the activity of enzymes which is restored when the samples are thawed. The formation of apocryphal protein spots in 2D gels indicates the inability to completely halt enzyme activity before protein fragmentation can occur. Further, alterations of the proteome caused by in vitro degradation may obscure important information of a sample's biological state and hinder the discovery of biomarkers. An alternative is to use rapid heat inactivation to eliminate enzymatic activity. The Denator Stabilizor T1 has been evaluated and shown to improve results by decreasing proteolytic fragments.

See more of this Session: Poster Session: Annual Meeting of the American Electrophoresis Society
See more of this Group/Topical: Topical 3: 2009 Annual Meeting of the American Electrophoresis Society (AES)

# VIABLE AND NON VIABLE MICROALGAE FRACTIONATION EMPLOYING INSULATOR-BASED DIELECTROPHORESIS

*Roberto C. Gallo-Villanueva, Nadia M. Jesús-Pérez, José I. Martínez-López, Adriana Pacheco-Moscoa, Mario M. Alvarez and Blanca H. Lapizco-Encinas**

*Tecnológico de Monterrey, Monterrey, NL, México*

## Abstract

Dielectrophoresis (DEP) is an electrokinetic transport mechanism driven by polarization effects on particles in the presence of non-uniform electric fields. Insulator-based DEP (iDEP) requires insulating structures to produce such non-uniform electric fields. iDEP is a non-destructive technique which can be effectively employed to fractionate and concentrate microorganisms. Differences in membrane's state and conductivity have an effect on the DEP response of cells. In this work, the potential of iDEP to fractionate live and dead microalgae (*Selenastrum capricornutum*) due to membrane integrity is demonstrated.

Heat-killed microalgae cells were prepared by heating a sample of live cells at 80 °C for five minutes, and mortality was assessed by staining with Sytox Green dye (Invitrogen), which can only penetrate damaged membranes. Live cells were visualized red (chlorophyll A autofluorescence). A mixture of viable and non-viable cells was then loaded at the inlet reservoir of a glass microchannel with an array of cylindrical insulating posts. Electrodes were placed at the inlet and outlet reservoirs in order to apply direct current (DC) electric fields. Dye excitation was done with UV light. Results demonstrated that microalgae cells exhibit negative dielectrophoretic behavior when exposed to DC electric fields, where viable cells experience a stronger dielectrophoretic response than non-viable cells. Differential negative dielectrophoretic trapping of live and dead cells was clearly obtained at an applied field of 800 V/cm. These results confirm the potential of this novel technique for rapid microalgae viability fractionation and assessment.

## Introduction

In the last decades, there has been increasing interest on the development of microscale separation systems. Miniaturization brings important advantages to separation technology, since a smaller sample volume allows for faster analysis and opens the possibility for more robust systems, reduced costs, greater sensitivity and portability.[1] There is a growing interest on the development of separation techniques with potential for miniaturization. Dielectrophoresis (DEP) has become one of the most important micro analytical techniques, since it can be applied for the manipulation, separation and concentration of macromolecules, virus, cell organelles, bacterial, yeast and parasites.[1-3]

Dielectrophoresis is a non-destructive, non-invasive electrokinetic transport mechanism, with great potential for the separation and manipulation of bioparticles in microfluidic devices. DEP was first described by Pohl[4,5] and it is defined as the movement of particles in a non-uniform electric field, due to polarizations effects. Particles having a higher polarizability than their immersion medium exhibit positive DEP: motion toward regions of greater field intensity. On the other hand, particles having a lower conductivity than their immersion medium exhibit negative DEP: motion away from regions of high electric field intensity. DEP can occur in alternating current (AC) or direct current (DC) electric fields. To produce the non-uniform fields required to carry out DEP, the most common approach is to apply AC electric fields employing arrays of microelectrodes. However, electrode-based DEP has important shortcomings, such as high cost of electrode construction, complex fabrication processes and decrease of functionality due to fouling effects, which is a common effect when handling biological samples.[6-10] There is an alternative manner of carrying out DEP; insulator-based-DEP (iDEP) is a technique where the electric field is applied employing only two electrodes that straddle an array of insulating structures. When an electric field is applied across an insulating structures array, the presence of the structures creates regions of higher and lower field strength, *i.e.* dielectrophoretic traps.[11,12] Insulator-based DEP systems do not lose their functionally despite of fouling effects, which makes them more suitable for biological applications; iDEP systems can be fabricated from a wide variety of materials, including plastics,[13-15] leading to inexpensive systems, increasing its potential for high throughput applications.

There are a number of studies that confirm the success of iDEP for the effective concentration and fractionation of biological particles non-destructively.[16-19] It has been demonstrated that differences in cell's membrane state and conductivity has an effect on the dielectrophoretic response of cells, allowing viability discrimination.[16] DEP has been proven useful to manipulate live and heat-treated cells of *Listeria* [6], viable and non-viable yeast cells [20] and live and dead bacteria.[16]

With regard to microalgae, the assessment of the changes in cell viability is of great significance for the monitoring of microalgae culturing processes for laboratory and semi-industrial scale. Several research groups have worked on methods to distinguish the viability of microalgae, like the absorption of elements like copper[21] or tributyltin.[22] However, a reliable viability assessment method to efficiently differentiate live from dead microalgae has not been developed yet. The titer of microalgae in culture is traditionally measured accurately based on reproducibility by viability assays in agar and liquid media. Nevertheless, both methods require media preparation and incubation for at least four days before titers can be evaluated.[23] Cell viability could also be reflected by changes in morphology, motility, membrane or physiological state.[24] Efforts made to define a reliable viability assessment method for algae include fluorescent microscopic methods,[23-25] chlorophyll auto-fluorescence; methods based on membrane permeability and MTT assays.[24] The study of cell dielectric properties as a function of cell physiological state and membrane conductivity can improve the rapid detection of the presence of viability of microorganisms,[26] including microalgae, thus eliminating the need for long procedures, reagents and expensive equipment. It is suggested that dielectrophoresis responses could be used as a direct mean to separate microalgae, therefore assessing viability of a sample according to viability reflected as changes in cell membrane.

In view of the need for a fast evaluation of the physiological state of algal cultures, the present work demonstrates the capability of iDEP to fractionate and concentrate viable and non-viable *Selenastrum capricornutum* cells. Results demonstrated that live and dead microalgae cells exhibit negative dielectrophoretic behavior when exposed to non uniform DC electric fields, where live cells exhibited a stronger dielectrophoretic response than dead cells.

## Theory of dielectrophoresis

The dielectrophoretic force acting on an isolated spherical particle can be represented as:

$$F_{DEP} = 2\pi\varepsilon_m r^3 \operatorname{Re}(f_{CM})\nabla E^2 \qquad (1)$$

where $\varepsilon_m$ is the permittivity of the suspending medium, *r* is the radius of the particle, *E* the local electric field and Re ($f_{CM}$) it is the real part of the Clausius-Mossotti (CM) factor. When low frequency electric fields are used (frequency $\leq$ 100 kHz), the CM factor can be approximated in terms of real conductivities as [27]:

$$f_{CM} = \left[\frac{\sigma_p - \sigma_m}{\sigma_p + 2\sigma_m}\right] \qquad (2)$$

From Equation (1) and (2) it is possible to observe that the dielectrophoretic force exerted on a particle can be positive or negative depending on the sign of the CM factor. The dielectrophoretic force is of second order with respect to the applied electric fields (Equation 1); at low applied electric fields, low electric fields gradients are produced, leading to negligible dielectrophoretic force.

Therefore, the applied electric field has to be high enough for DEP to become significant and immobilize particles.

When iDEP is applied employing DC electric fields, other electrokinetic transport mechanisms are present: electrophoresis (EP) and electroosmotic flow (EOF).[13] Electrokinetics (EK) is the superposition of electrophoresis and electroosmosis. According to that, there are two regimes of iDEP: streaming and trapping DEP. Streaming DEP occurs when the dielectrophoretic force overcomes diffusion but does not overcome the EK flow. Trapping DEP is the reversible immobilization of particles in dielectrophoretic traps and it occurs when the DEP force overcomes diffusion and EK flow (as well as pressure-driven flow if present).[11,19,28]

EOF is the movement of liquid when a DC electric field is applied in a microchannel. For electroosmosis to occur it is necessary that the channel walls posses a net charge. In the present work, microdevices made from glass were employed, and the glass surface has a net negative charge in the presence of an electrolyte. The ions from the electrolyte will form a double layer over the glass surface. When a DC electric field is applied, the ions from the double layer are attracted to the negative electrode, generating liquid flow. The EOF velocity is defined as [26]:

$$v_{EO} = \mu_{EO} E \quad (4)$$

where $v_{EO}$ is the EOF velocity, $\mu_{EO}$ is the electroosmotic mobility and $E$ is the applied electric field. Electrophoresis is the movement of charged particles in the presence of an electric field; the electrophoretic velocity of a particle is defined as:

$$v_{EP} = \mu_{EP} E \quad (5)$$

where $v_{EP}$ is the electrophoretic velocity, $\mu_{EP}$ is the electrophoretic mobility and $E$ is the applied electric field. The sum of both electroosmosis and electrophoresis on particle motion relative to a fixed channel can be described by an electrokinetic mobility. For this glass system the electrokinetic velocity is obtained employing the electrokinetic mobility as follows:

$$v_{EK} = \mu_{EK} E = (\mu_{EO} - \mu_{EP}) E \quad (6)$$

where $v_{EK}$ is the electrokinetic velocity, $\mu_{EK}$ is the electrokinetic mobility and $E$ is the applied electric field.

The dielectrophoretic velocity, which is a second order effect of the electric field, for a negative DEP particle, can be expressed as:[29]

$$v_{DEP} = -\mu_{DEP} \nabla E^2 \quad (7)$$

where $v_{DEP}$ is the dielectrophoretic velocity, $\mu_{DEP}$ is the dielectrophoretic mobility and $E$ is the applied electric field. In order to achieve dielectrophoretic trapping, the dielectrophoretic velocity has to be greater than the EK velocity and the effect of diffusion and pressure-driven flow if present. This research work is focused on the demonstration of live and dead microalgae fractionation employing trapping iDEP.

## Materials and Methods

### Sample preparation

A green algae strain, *Selenastrum capricornutum* (UTEX 1648), was grown at 24° C in Modified Bold3N medium in 50 mL Falcon tubes under continuous illumination by a fluorescent lamp (100 μmolE $m^{-2}$ $s^{-1}$) and air conditions containing 0.025 v/v $CO_2$. A sample of 1 mL mid-logarithmic phase culture of microalgae was divided into two equal parts, which were centrifuged at 3000 rpm for 3 minutes. Both pellets were resuspended in 0.5 mL distilled water. One of them was used to prepare dead microalgae by heat treatment, in water bath at 80° C for 5 minutes.

Mortality of heat-treated cells was assessed by staining with Sytox Green dye, excitation/emission 504/523 nm (S7020, Invitrogen, Carlsbad CA, USA), which can only penetrate damaged membranes. The SYTOX Green fluorescent dye was supplied as a 5 mM stock solution in DMSO. 0.5 μl of this stock solution was added to 0.5 μl cell suspension for a final dye concentration of 5 μM, and the mixture was incubated for 30 minutes at room temperature in the dark. No washing was required. The non-treated sample was not stained since chlorophyll A, which is presented inside the cells and fluoresce red under UV light, will be used to distinguish from the heat-treated sample. The cell suspensions were then mixed to produce samples containing 50% live cells.

**Equipment**

Fluorescence of SYTOX Green and chlorophyll A fluorescence were observed simultaneously, allowing the discrimination of viable and non-viable cells. Image capturing was done employing an inverted microfluidic microscope Axiovert 200 with an integrated camera (Carl Zeiss, Göttingen, Germany). A highvoltage power supply HVS448 (LabSmith, Livermore, CA, USA) was also used to apply DC electric fields.

**Insulator-based DEP experiments**

A microchannel made from glass was employed, the channel was 10.16-mm-long, 1-mm-wide, 10-μm deep, and had an array of 32 cylindrical posts distributed in eight columns of four posts each, posts had a diameter of 200-μm and were arranged in a square array 250-μm center-to-center. The cylindrical posts transverse the entire height of the microchannel (Figure 1).

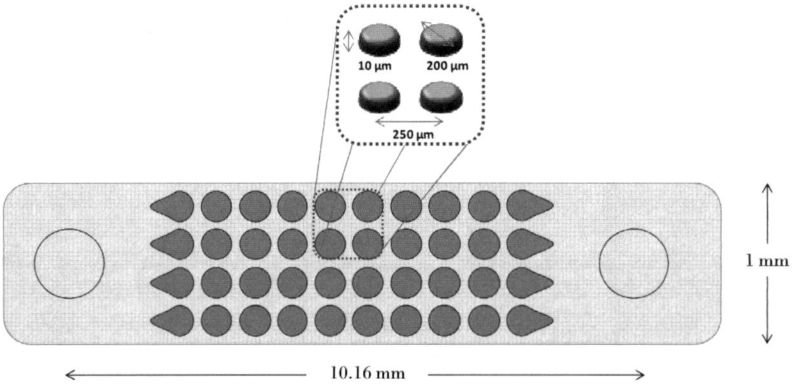


**Figure 1**. Schematic representation of the microchannel employed with iDEP experimentation.

A mixture of viable and non-viable cells was then loaded at the inlet reservoir of the microchannel containing the array of cylindrical insulating posts. Care was taken to eliminate pressure-driven flow. Platinum electrodes were placed at the inlet and outlet reservoirs in order to apply direct current (DC) electric fields.

**Results and Discussion**

According to Equation (6) electrokinetic mobility depends linearly to the electric field applied, while Equation (7) shows that the dielectrophoretic mobility is a non-linear relation to such field. This means that under low applied electric fields the former will dominate the system in the form of electroosmotic flow, but as the field is increased DEP will overcome EOF and particles will be dielectrophoretically trapped. In this case, it is expected that under DC fields, bioparticles exhibit negative dielectrophoretic behavior,[16] *i.e.,* particles will be repulsed from the areas of higher field intensity (narrow regions between the cylindrical posts). The magnitude of the rejection will be determined by Equation (1) which includes the CM factor, also called "polarization factor," which is the relationship between the particle and the medium conductivities. This means that two particles of the same size, but with different conductivities, will experience dielectrophoretic forces of different magnitudes.

Viable and non-viable, non-treated and heat-treated samples respectively, will present differences in their membranes which will be reflected on a change in their conductivities; hence their DEP response will also vary. When a cells dies, its low conductive membrane becomes compromised, and it gets contaminated with cells cytoplasm which is very conductive, leading to a significant increase on the conductivity of the membrane.[16] Therefore, according to Equation (2), viable algae will experience more negative dielectrophoresis compared to non-viable algae, *i.e.* live algae will exhibit a stronger repulsion from the high gradient electric field region (constriction between cylindrical posts).

Figure 2 shows a sequence of a mixture of 50% of viable and non-viable cells under electric fields of 0, 500 and 800 V/cm inside the microchannel described in Figure 1. Figure 2a shows the sample prior to the application of the electric field, where live (red) and dead (green) cells can be clearly visualized and differentiated. Figure 2b shows that to an applied field of 500 V/cm, dead cells (green) are dominated by EOF and are flowing through the post array; while live cells (red) exhibit some negative dielectrophoretic immobilization forming loose bands of trapped cells just prior to each constriction between the cylindrical posts. Furthermore, as the field is increased to 800 V/cm (Fig. 2c) both cell types are immobilized, and viable cells experience stronger repulsion from the region of higher field intensity (narrow part between posts) than non-viable cells and are trapped in different regions. This is observed as two different bands in Figure 2c.

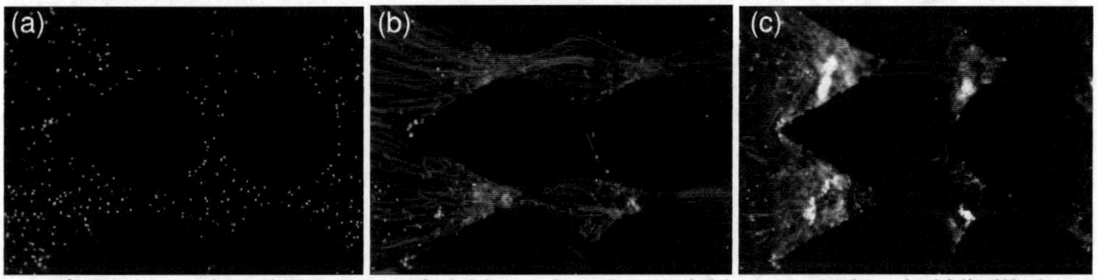


**Figure 2.** Dielectrophoretic response of viable (red) and non-viable (green) algae in bidistilled water. (a) Under 0 V/cm cells are seen static; (b) with a field of 500 V/cm EOF dominates dead cells with some dielectrophoretic trapping of live cells; (c) by increasing the electric field to 800 V/cm strong DEP trapping is presented and a separation of dielectrophoretic bands is seen, viable cells experience stronger repulsion as the non-viable cells.

## Conclusions

The present study demonstrates manipulation and separation between viable and non-viable algae (*Selenastrum capricornutum*) employing iDEP inside a microchannel manufactured in glass. DC fields were applied; therefore EOF was present, eliminating the need of a micropump. Results show that live and dead cells exhibit different dielectrophoretic responses. At a field of 500 V/cm mainly live cells are dielectrophoretically immobilized while dead cells can flow through. By increasing the electric field to 800 V/cm it was possible to trap both types of cells obtaining spatial separation. The dielectrophoretic discrimination of live and dead cells was possible due to their differences in membrane integrity, which is reflected in conductivity differences between live and dead cells.

These results confirm the potential of this novel technique for rapid microalgae viability fractionation and assessment with great potential as cell-viability method and as a one-step separation and concentration technique. Further investigation in the field can bring enormous benefits to algae cultivation and study.

## Acknowledgments

The authors would like to acknowledge the financial support provided by the grant CONACYT-CB-2006-53603. The authors are grateful for the financial support provided by *Cátedras de Investigación* (CAT122 and CAT142) of *Tecnológico de Monterrey.*

## References

1 Lapizco-Encinas B.H., Rito-Palomares M., (2007), "Dielectrophoresis for the manipulation of nanobioparticles," *Electrophoresis* 28(24), pp.4521-4538.

2 Gascoyne P.R.C., Vykoukal J., (2002), "Particle separation by dielectrophoresis," *Electrophoresis* 23(13), pp.1973-1983.

3 Voldman J., (2006), "Electrical forces for microscale cell manipulation," *Ann. Rev. Biomed. Eng.* 8, pp.425-454.

4 Pohl H.A., (1951), "The Motion and Precipitation of Suspensoids in Divergent Electric Fields," *J. Appl. Phys.* 22(7), pp.869-871.

5 Pohl H.A., (1978), "Dielectrophoresis," Cambridge University Press, Cambridge.

6 Li H., Bashir R., (2002), "Dielectrophoretic separation and manipulation of live and heat-treated cells of Listeria on microfabricated devices with interdigitated electrodes," *Sens. Actuator B-Chem.* 86(2-3), pp.215-221.

7 Medoro G., Manaresi N., Altomare L., Leonardi A., Tartagni M., Guerrieri R., "A Lab-on-a-chip for Cell Separation Based on the Moving-Cages Approach," (2002), in 16th European Conference on Solid-State Transducers (Eurosensors XVI), Proc. Eurosensors XVI, Prague, Czech Republic, vol. 47-51.

8 Markx G.H., Huang Y., Zhou X.F., Pethig R., (1994), "Dielectrophoretic characterization and separation of microorganisms," *Microbiology-UK* 140, pp.585-591.

9 Rousselet J., Markx G.H., Pethig R., (1998), "Separation of erythrocytes and latex beads by dielectrophoretic levitation and hyperlayer field-flow fractionation.," *Colloids and Surfaces A: Physiochem. Eng. Aspects* 140(1-3), pp.209-216.

10 Washizu M., Kurosawa O., (1990), "Electrostatic manipulation of DNA in microfabricated structures," *IEEE Trans. Ind. Appl.* 26(6), pp.1165-1172.

11 Cummings E.B., Singh A.K., (2003), "Dielectrophoresis in Microchips Containing Arrays of Insulating Posts: Theoretical and Experimental Results," *Anal. Chem.* 75(18), pp.4724-4731.

12 Kang Y., Li D., Kalams S., Eid J., (2008), "DC-Dielectrophoretic separation of biological cells by size," *Biomed. Microdev.* 10(2), pp.243-249.

13 Davalos R.V., McGraw G.J., Wallow T.I., Morales A.M., Krafcik K.L., Cummings E.B., Simmons B.A., (2008), "Performance impact of dynamic surface coatings on polymeric insulator-based dielectrophoretic particle separators," *Anal. Bioanal. Chem.* 390(3), pp.847-855.

14 Sabounchi P., Morales A.M., Ponce P., Lee L.P., Simmons B.A., Davalos R., (2008), "Sample concentration and impedance detection on a microfluidic polymer chip.," *Biomed. Microdev.* 10(5), pp.661-670.

15 Simmons B.A., McGraw G.J., Davalos R.V., Fiechtner G.J., Fintschenko Y., Cummings E.B., (2006), "The Development of Polymeric Devices as Dielectrophoretic Separators and Concentrators," *MRS Bulletin* 31, pp.120-124.

16 Lapizco-Encinas B.H., Simmons B.A., Cummings E.B., Fintschenko Y., (2004), "Dielectrophoretic Concentration and Separation of Live and Dead Bacteria in an Array of Insulators," *Anal. Chem.* 76(6), pp.1571-1579.

17 Lapizco-Encinas B.H., Simmons B.A., Cummings E.B., Fintschenko Y., (2004), "Insulator-Based Dielectrophoresis for the Selective Concentration and Separation of Live Bacteria in Water," *Electrophoresis* 25, pp.1695-1704.

18 Lapizco-Encinas B.H., Davalos R., Simmons B.A., Cummings E.B., Fintschenko Y., (2005), "An Insulator-Based (Electrodeless) Dielectrophoretic Concentrator for Microbes in Water," *J. Microbiol. Methods* 62, pp.317-326.

19 Ozuna-Chacón S., Lapizco-Encinas B.H., Rito-Palomares M., Martínez-Chapa S.O., Reyes-Betanzo C., (2008), "Performance characterization of an insulator-based dielectrophoretic microdevice," *Electrophoresis* 29(15), pp.3115-3122.

20 Markx G.H., Talary M.S., Pethig R., (1994), "Separation of viable and nonviable yeast using dielectrophoresis," *J. Biotechnol.* 32(1), pp.29-37.

21 Al-fawwaz A.T., Wan-Maznah W.O., "Biosorption of Copper Using Live and Dead Green Microalgae Isolated form Penang Rivers, Malasya," *International Conference on Enviromental Research and Technology, Penang, Malasya* Proc, pp.940-943.

22 Tam N.F.Y., Chong M.Y., Wong Y.S., "Removal of tributylin (TBT) by live and dead microalgal cells," *Marine Pollution Bulletin* 45, pp.362-371.

23 Sato M., Murata Y., Mizusawa M., Iwahashi H., Oka S., (2004), "A Simple and Rapid Dual-fluorescence Viability Assay for Microalgae," *Microbiol. Cult. Coll.* 20(2), pp.53-59.

24 Breeuwer P., Abee T., (2000), "Assessment of viability of microorganisms employing fluorescence techniques," *International Journal of Food Microbiology* 55, pp.193-200.

25 Pouneva I., (1997), "Evaluation of Algal Culture Viability and Physiological State by Fluorescent Microscopic Methods," *Bulg. J. Plant Physiol* 23(1-2), pp.67-76.

26 Wu Y., Huangb C., Wang L., Miaob X., Xing W., Cheng J., (2005), "Electrokinetic system to determine differences of electrorotation and traveling-wave electrophoresis between autotrophic and heterotrophic algal cells," *Colloids and Surfaces A: Physicochem. Eng. Aspects* 262, pp.57-64.

27 Markx G.H., Dyda P.A., Pethig R., (1996), "Dielectrophoretic separation of bacteria using a conductivity gradient," *J. Biotechnol.* 51(2), pp.175-180.

28 Cummings E.B., Singh A.K., "Dielectrophoretic trapping without embedded electrodes," (2000), in Mastrangelo CH, Becker H (Editors), Proceedings of SPIE: Conference on Microfluidic Devices and Systems III, Santa Clara, CA, vol. 4177, pp. 164-173.

29 Cummings E.B., "A Comparison of Theoretical and Experimental Electrokinetic and Dielectrophoretic Flow Fields," (2002), in 32nd AIAA Fluid Dynamics Conference and Exhibit 2002, American Institute of Aeronautics and Astronautics, St. Louis, Missouri, vol. paper 2002-3193, pp. 1-17.

# Pressure Enhanced Processes Enabling High Quality Two-Dimensional Gel Electrophoresis of *Frankia* Mycelia and Vesicle Structures*

*Louis S. Tisa, University of New Hampshire, Durham, NH; Gary B. Smejkal, Harvard Medical School, Boston, MA; Thomas Berkelman, BioRad Laboratories, Hercules, CA; and Winston P. Kuo, Harvard Medical School, Boston, MA.*

## Abstract

The actinobacteria, *Frankia*, are particularly challenging in terms of sample preparation, isoelectric focusing (IEF) and two-dimensional gel electrophoresis (2DE). While French Press is typically used for the disruption of *Frankia* mycelia, protein yields are typically very low and biased toward cytoplasmic proteins. Operating at a maximum pressure of 20,000 psi, the French press fails to disrupt vesicles such that this recalcitrant organelle has remained largely uncharacterized in terms of its protein constituency. Alternatively, a pressure cycling technology (PCT) that recycles the sample at 45,000 psi maximum pressure effectively disrupts both hyphae and the resilient vesicles. However, increased protein yield from *Frankia* strain EAN1pec corresponds to increased release of an indigenous red pigment which interferes with protein assay and IEF. The pigment is lessened, but not removed by acetone precipitation and concentrates with proteins in ultrafiltrative retentates, suggesting the pigment exists as high molecular weight aggregates which are not dissociated chaotropes, detergents or reducing agents. If not removed, the pigment precipitates on the surface of the immobilized pH gradient (IPG) and prevents proteins from being fully imbibed into the strip. This is evidenced by the preponderance of horizontal streaking in second dimension gels. Pigment that is imbibed in the IPG rapidly focuses in the pH 4-5 region of the gradient, creating a highly conductive zone where a localized voltage drop prevents proteins from properly focusing. Hence, it is ambiguous as to whether proteins that are aligned with this zone are due to their association with the pigment or as an electrophoretic phenomenon. Ultrafiltration of mycelia and vesicle homogenates with porous membranes of 100 kDa nominal molecular weight limit partitioned 80% of the initial sample volume containing approximately 60% of the protein mass into the filtrate, whereas most of the offending pigment was accumulated in the retentate. Consequently, over 500 proteins were resolved in two-dimensional gels of vesicle filtrates.

* Poster presentation at the 2009 American Electrophoresis Society Meeting, Nashville, TN, USA, November 12-14, 2009.

## Introduction

*Frankia* infect roots and induce nodule formation in a variety of woody dicotyledonous plants, leading to a symbiotic nitrogen-fixing association [1, 2]. These bacteria are developmentally complex forming three cell types: vegetative hyphae, spores located in sporangia, and unique lipid-enveloped organelles called vesicles. Vesicles are formed inside of the plant cells of the nodule or in culture under nitrogen limiting conditions and act as specialized structures for the nitrogen fixation process [3]. Their shape is strain-dependent and host-plant-influenced. Vesicles are formed terminally on short side branches of hyphae and have a septum near their base. The mature vesicle is surrounded by an envelope that extends down the stalk of the vesicle past the basal septum which separates the vesicle from the hyphae. One of the functions of the vesicle is to protect the nitrogenase enzyme from oxygen inactivation.

Methods have been described for the isolation and purification of intact vesicles from *Frankia* grown in culture [4, 5]. These purified vesicles retain nitrogenase activity. Initial investigations on the physiological properties of purified vesicles have focused on nitrogen and carbon metabolism, and bioenergetics [6, 7]. Vesicles are very resilient structures that are difficult to disrupt and protein extraction from vesicles has historically been difficult and a significant bottleneck to proteomic approaches. In these studies, maximal protein recoveries were attained using a pressure cycling technology (PCT) that recycled the samples at 45,000 psi maximum pressure in combination with optimized chemistries. Interference with IEF of proteins was minimized in 100 kDa filtrates enabling unprecedented numbers of proteins to be separated by 2DE

## Materials and Methods

### *Growth conditions*

*Frankia* strain EAN1pec [8] was grown and maintained in basal growth medium with $NH_4Cl$ as the nitrogen source, as described previously [9]. Large scale batch cultures of strain EAN1pec were obtained by growing cells in a carboy with 15 liters of medium containing 20 mM succinate or 20 mM fructose with limited 0.5 mM $NH_4Cl$ as described previously [5]. Under these conditions, the cells depleted their supply of $NH_4Cl$ after 7 to 8 days of growth and were growing with $N_2$ as the nitrogen source when harvested.

### *Vesicle isolation and purification*

Vesicles were isolated and purified as described previously [5, 7]. Freshly harvested or frozen cells were washed twice with 25 mM Tris-HCl, 0.5 M mannitol buffer pH 7.4 at 20°C. Vesicles were isolated by passing the washed culture through a French pressure cell at 10,000-20,000 psi at 4°C. Prior to passage through the French pressure cell, protease inhibitor cocktail (Sigma-Aldrich Chemical, St, Louis, MO, USA) was added to the cell suspension to prevent protein degradation. As a result of this treatment the mycelia are completely disrupted but the vesicles remain intact. Vesicles were purified from the cytosolic and membrane fractions by a series of low speed centrifugations at 20°C. The numbers of vesicles were counted by using a Petroff-Hausser counting chamber with a phase-contrast microscope at a 400X magnification.

### *Protein Extraction*

Proteins were extracted by PCT as described [10, 11] using reagents from the ProteoSOLVE CE Stringent and SB Kits (Pressure BioSciences, South Easton, MA, USA). For this, mycelium and purified vesicle suspensions were collected by centrifugation at 10,000 x g for 10 minutes and 100 mg of each pellet was transferred to a PULSE Tube. One hundred milligrams of CE SiC abrasive from the ProteoSOLVE CE kit and 50 uL of concentrated protease inhibitor cocktail from Sigma-Aldrich Chemical (St, Louis, MO, USA) were added and the samples were mixed thoroughly and flash frozen. This frozen amalgam impregnated with the abrasives was then extruded through the perforated lysis disk of the PULSE Tube, followed by the addition of 1200 uL of IEF Reagent from the ProteoteoSOLVE SB Kit and 35 uL of 200 mM tributylphosphine. Following 30 pressure cycles at 35,000 psi maximum pressure, the PULSE Tubes were coupled to the insert of an Ultrafree CL microfilter (Millipore Corporation, Danvers, MA, USA) and evacuated by centrifugation at 500 RCF for two minutes. The filtrates were alkylated by the addition of 10 mM acrylamide monomer. The alkylation reaction was quenched after two hours by the addition of 50 mM dithiothreitol [12]. The samples were concentrated using an Ultrafree centrifugal ultrafiltration device with 10 kDa nominal molecular weight limit (NMWL). Protein concentrations of the retentates was determined by Bradford assay (BioRad Laboratories, Hercules, CA, USA).

***Two-Dimensional Polyacrylamide Gel Electrophoresis (2GE)***

For electrophoresis, the samples were further clarified using an Ultrafree centrifugal ultrafiltration device with 100 kDa NMWL and the filtrates reserved for analysis [13]. Immobilized pH gradients (IPGs) spanning pH 4-7 were hydrated with 200 uL of each sample for six hours prior isoelectric focusing for 100,000 Volthours at 24°C. Prior to second dimension PAGE, the IPGs were equilibrated twice for ten minutes in 375 mM Tris HCl pH 8.8 containing 2.5% SDS, 50 mM DTT, 0.005% bromophenol blue, and 0.005% phenol red. Second dimension PAGE was performed on 8-16% gradient gels (BioRad Laboratories, Hercules, CA, USA). Gels were fixed overnight in 25% ethanol, 10% acetic acid and silver stained [14].

## Results

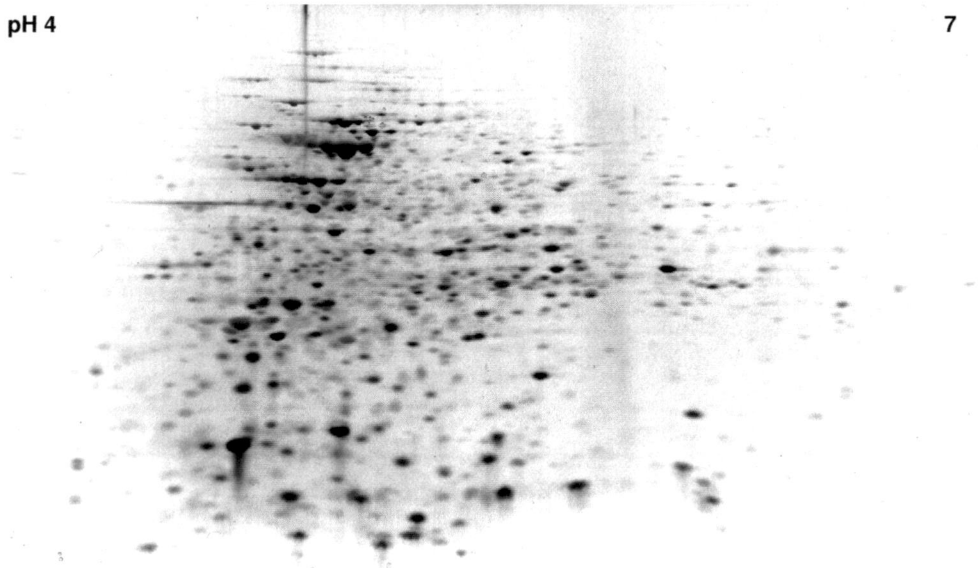


**Figure 1.** 2DE of 100 kDa filtrate of *Frankia* $NH_4Cl$ whole cell lysate. IPGs were pH 4-7 range. Second dimension PAGE was 8-16% linear polyacrylamide gradient.

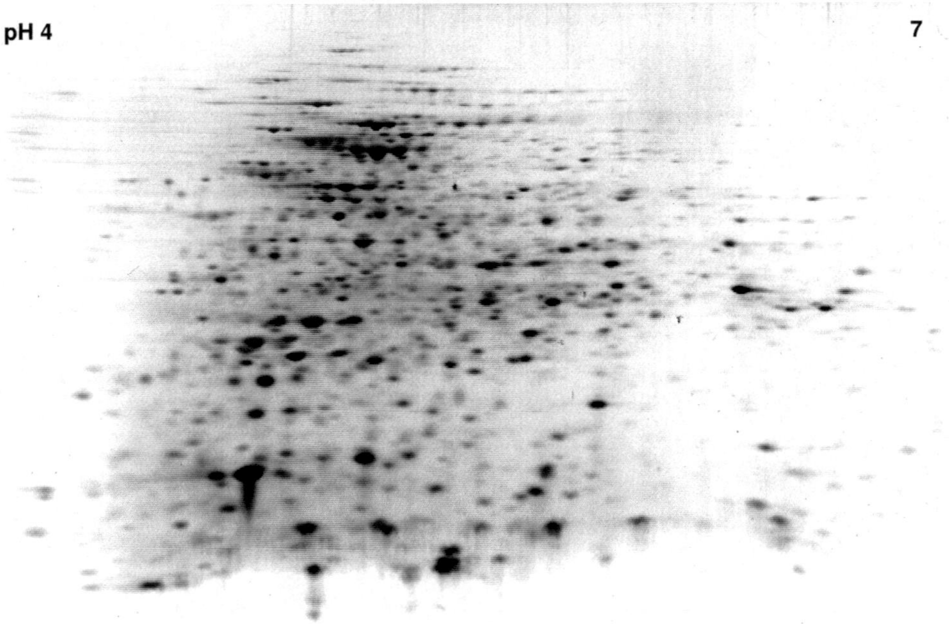


**Figure 2.** 2DE of 100 kDa filtrate of *Frankia* $N_2$ whole cell lysate. IPGs were pH 4-7 range. Second dimension PAGE was 8-16% linear polyacrylamide gradient.

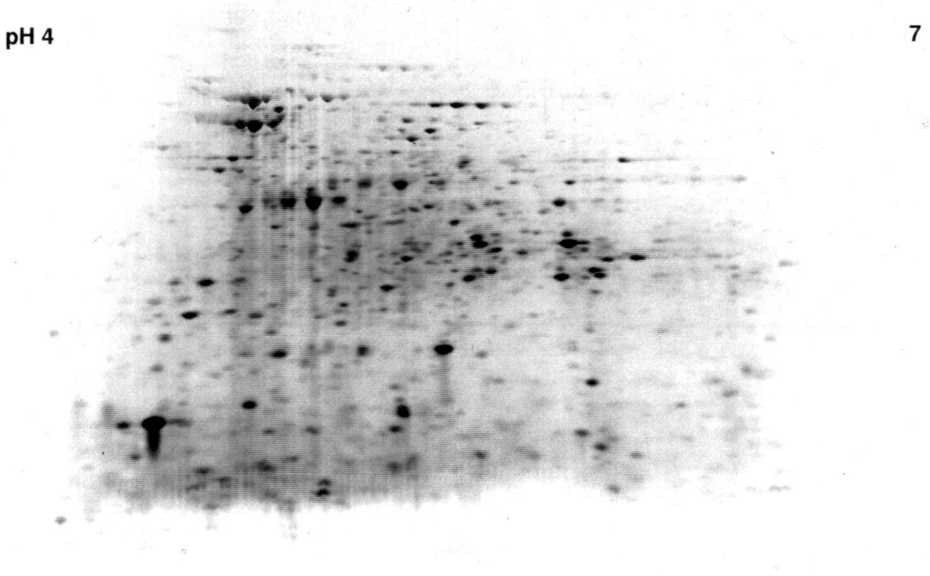


**Figure 3.** 2DE of 100 kDa filtrate of *Frankia* vesicle fraction. IPGs were pH 4-7 range. Second dimension PAGE was 8-16% linear polyacrylamide gradient.

**Acknowledgements:**

LST was funded by NSF Grant No. EF-0333177. GBS is Affiliate Assistant Professor at UNH.

**Rerferences:**

[1] Benson, D.R., and Silvester, W.B. 1993. Biology of *Frankia* strains, actinomycete symbionts of actinorhizal plants. *Microbiological Review* 57:293-319.

[2] Wall, L.G. 2000. The actinorhizal symbiosis. *Journal of Plant Growth and Regulation* 19:167-182.

[3] Silvester, W.B., R.H. Berg, C.R. Schwintzer, and J.D. Tjepkema. 2008. Oxygen responses, hemogloblin, and the structure and function of vesicles. In: *Nitrogen-fixing Actinorhizal Symbioses*. Pawlowski, K. and Newton, W.E., eds. Springer, Dordrecht, The Netherlands, pp.105-146.

[4] Noridge, N.A. and Benson, D.A. 1986. Isolation and nitrogen-fixing activity of *Frankia* sp. strain CpI1 vesicles. *Journal of Bacteriology* 166:301-305

[5] Tisa, L. S. and Ensign, J. C. 1987. Isolation and nitrogenase activity of vesicles from *Frankia* sp. strain $EAN1_{pec}$. *Journal of Bacteriology* 169:5054-5059.

[6] Schultz, N.A. and Benson, D. R. 1989. Enzymes of ammonia assimilation in hyphae and vesicles of *Frankia* sp. strain CpI1. *Journal of Bacteriology* 172:1380-1384.

[7] Tisa, L.S. and Richards, J. 2004. Metabolic capacity of purified vesicles from *Frankia* EAN1pec. *Symbiosis* 37:365-377.

[8] Lalonde, M., Calvert, H.E., and Pine, S. 1981. Isolation and use of *Frankia* strains in actinorhizae formation. In: *Current perspectives in nitrogen fixation*. Gibson, A.H. and Newton, W.E., eds. Australia Academy of Science, Canberra, Australia, pp. 296-299.

[9] Tisa, L.S., Chval, M.S., Krumholz, G.D., and Richards, J. 1999. Antibiotic resistance patterns of *Frankia* strains. *Canadian Journal of Botany*. 77:1257-1260.

[10] Ringham, H., Pedrick, N., Smejkal, G.B., Behnke, J., and Witzmann, F.A. 2007. Application of pressure cycling technology to tissue sample preparation for 2-DE. *Electrophoresis* 28:1022-1024.

[11] Smejkal, G., Witzmann, F.A., Ringham, H., Small, D., Chase, S., Behnke, J., and Ting, E. 2007. Sample Preparation for Two-Dimensional Gel Electrophoresis using Pressure Cycling Technology. *Analytical Biochemistry* 363:309-311.

[12] Smejkal, G.B., Li, C., Robinson, M.H., Lazarev, A., Lawrence, N., and Chernokalskaya, E. 2006. Simultaneous Reduction and Alkylation of Protein Disulfides in a Centrifugal Ultrafiltration Device Prior to Two-Dimensional Gel Electrophoresis. *Journal of Proteomic Research* 5: 983- 987.

[13] Smejkal, G.B., Kwan, A,T., Zianni, M., Mandich, D., Mason, M., and Koch, J. 2008. Pressure Cycling Technology mediated sample preparation schemes enabling high quality two-dimensional gel electrophoresis of *Fagus grandifolia* tree bark proteins. *Plant and Animal Genome XVI Conference*, San Diego, CA, January 12-16, 2008.

[14] Smejkal, G.B. 2005. Protein Staining in Polyacrylamide Gels. *Separation Methods in Proteomics,* CRC Press, Taylor & Francis, Boca Raton, pp 439-452.

# Microfluidic Cell Culture Device to Study the Effect of Spatial Confinement On Adipocyte Cell Death Rates

**Sachidevi Puttaswamy**, Biological Engineering, University of Missouri, Columbia, MO

Nawar Faruque, Biological Engineering, University of Missouri, Columbia, MO

James Perfield, Nutrition & Exercise Physiology, University of Missouri, Columbia, MO

Shramik Sengupta, Biological Engineering, University of Missouri, Columbia, MO

Being obese greatly increases one's risk of being afflicted by chronic diseases like diabetes, cardiovascular disease and stroke. These adverse metabolic states are believed to arise due to the secretion of various inflammatory factors originating from adipose (fat) tissue. Recent studies have indicated that with obesity, adipose cells expand so rapidly that the extracellular matrix surrounding them cannot expand at the same rate. This physical confinement results in increased cell death and macrophage infiltration. These events lead to a chronic state of inflammation in adipose tissue which spills over systemically to other tissues leading to the adverse metabolic conditions mentioned above.

With standard cell culture petridishes, it is not been possible to systematically investigate the effect of physical confinement on adipocyte death, and in vivo models have a large number of parameters that cannot be independently controlled. We present here the a "microfluidic" cell culture "vessel" embedded in a 3" x 1" cassette that consists of two types of pores: larger pores (50-200 microns) into which adipocytes are packed, and smaller pores (2-5 microns) that act like a vascular bed through which nutrients can be provided. Controlling the size of the larger pores provides us with a way to confine adipocyte growth to varying (but known) degrees. This system enables us to study the effect of spatial confinement in isolation from other Biomolecular factors on the death rate of adipose cells from various sources (subcutaneous, intra-abdominal etc.)

See more of this Session: Poster Session: Annual Meeting of the American Electrophoresis Society
See more of this Group/Topical: Topical 3: 2009 Annual Meeting of the American Electrophoresis Society (AES)

# Magnifying Differences in the Cross Over Frequency of Bacteria Using Solvent Modification and Antibody Tagged Nanoparticles

**Neha Gujarati**, Biological Engineering, University of Missouri-Columbia, Columbia, MO

Jinwang Tan, Biological Engineering, University of Missouri-Columbia, Columbia, MO

Luis Polo-Parada, Medical Pharmacology and Physiology, University of Missouri-Columbia, Columbia, MO

Shramik Sengupta, Biological Engineering, University of Missouri-Columbia, Columbia, MO

ielectrophoresis (DEP) can potentially serve as a means to rapidly isolate pathogenic bacterial species and strains of interest from a mixture containing other related, often non-pathogenic, species. Two species can be separated from each other easily using DEP if they have different Cross-Over Frequencies (COFs). The COF depends of a bacterial cell in a given solution depends on the properties (conductivity and dielectric constant) of the cell membrane and the cell interior, and on those of the solution in which it is suspended, and thus each species should have a different characteristic cross-over frequency.

In practice, however, it is seen that these crossover frequencies of bacteria can be very close together, thus making their separation using DEP quite difficult and inefficient. We propose a practical solution to this problem by formulating a "DEP buffer" consisting of Mannitol, Amino Hexanoic Acid, and Sodium Chloride in which the differences in COF are magnified. The buffer is hypertonic, has a low conductivity but high dielectric constant, and by magnifying the contribution of the properties of the cell membrane in determining the COF, enables resolution of many different bacterial species like E.coli, Lactobacillus, Pseudomonas, and Salmonella.

Closely related strains (such as E coli O157 and E coli K-12) however, are still not resolvable. We present a way of doing so using antibody-tagged nano-particles to bind to particular strains. Once these particles bind to the bacteria, the COF of the nano-particle bacteria conjugate differs from that of the bacteria alone.

These practical techniques open the door to detecting the presence of target pathogenic bacteria from mixtures containing large numbers of other bacteria

See more of this Session: Poster Session: Annual Meeting of the American Electrophoresis Society

See more of this Group/Topical: Topical 3: 2009 Annual Meeting of the American Electrophoresis Society (AES)

# Monitoring Bioremediation Using Single Strand Conformational Polymorphism and Capillary Electrophoresis

**Alice C. Jernigan**, Chemical Engineering, University of Arkansas, Fayetteville, AR

Duane C. Wolf, Dept of Crop,Soil and Environmental Sciences, University of Arkansas, Fayetteville, AR

Greg Thoma, Cheg, University of Arkansas, Fayetteville, AR

Christa Hestekin, Cheg, University of Arkansas, Fayetteville, AR

Bioremediation is the process by which microbes are used to break down soil contaminants. The traditional method for looking at bioremediation is by viable plate counts or microscopic examination. These methods are both time consuming and can underestimate the numbers due to the inability to culture many soil organisms. The increased speed and automation of capillary electrophoresis (CE) allows for more rapid and reliable results. The identification and assessment of activity of microbial soil communities can be accomplished by isolating DNA and RNA from soil samples. These DNA/RNA samples can then be characterized by a technique called single-strand conformational polymorphism (SSCP) analysis of the 16S gene and its variable regions. These variable regions have been amplified by polymerase chain reaction (PCR) giving species-dependent fragments of the gene sequence. SSCP, a method that is normally used to detect mutations, is the heat denaturing of these fragments resulting in single strand DNA (ssDNA). Snap cooling of the ssDNA causes it to refold into different secondary structures (conformations) depending on the nucleotide sequence. These different conformations can be separated due to their different electrophoretic mobility, even if they are the same size. CE-SSCP has been used to identify mixed populations from DNA extracted from soil samples and to identify activity of the microorganisms by examining the products of reverse transcription (RT)-PCR which relates to the sample's RNA. By isolating both DNA and RNA from the same soil sample, at the same time, a more accurate picture of the microbial community in those samples has been made available. This is an important development in the study of contaminated soils and the monitoring of the progress of bioremediation.

See more of this Session: Poster Session: Annual Meeting of the American Electrophoresis Society
See more of this Group/Topical: Topical 3: 2009 Annual Meeting of the American Electrophoresis Society (AES)

# PARTICLE MANIPULATION EMPLOYING ALTERNATING CURRENT ELECTRIC FIELDS IN AN ARRAY OF INSULATORS

**Javier L. Baylon-Cardiel, Ana V. Chávez-Santoscoy, Sergio O. Martínez-Chapa, and Blanca H. Lapizco-Encinas***

*Tecnológico de Monterrey, Monterrey, NL, México.*

## Abstract

Applying an electric potential that varies in time, microparticle manipulation is achieved in a microchannel with insulating structures. Particles are trapped employing insulator-based dielectrophoresis. The location of trapped particles shows a time dependency, which is studied varying the frequency of the applied electric potential.

## Introduction

Dielectrophoresis (DEP) is the motion of particles subject to a non-uniform electric field due to the effects of the induced polarization. This phenomenon was first described by Pohl in 1951,[1] and it has been developed as a useful electrokinetic technique for particle manipulation. In this regard, the DEP technique has been successfully employed for the transport and concentration of a wide variety of particles, ranging from proteins to microorganisms.[2-4]

Dielectrophoresis can occur in either alternating current (AC) or direct current (DC) electric fields. Traditionally, studies on DEP have employed AC electric field and arrays of metallic electrodes, where numerous geometries and configurations have been tested successfully. The manipulation of particles with such arrays of electrodes depends greatly on the frequency of the AC field, and it is achieved for large values of the frequency. [3-6]

On the other hand, insulator-based DEP (iDEP) has been typically carried out using DC electric fields.[7,8] In this approach, only two electrodes are employed, each located at the entrances of a microfluidic channel. Insulating structures are distributed along the channel, and a non-uniform electric field is generated between the structures when an electric potential is applied. This approach for DEP was first presented by Cummings and Singh.[9]. In their work, they described a device consisting of an array of cylindrical insulating posts located along a microfluidic channel.

The combination of AC electric fields with iDEP has been reported by Chou.[10] In their work, concentration of DNA is achieved using a microfluidic channel with quadrilateral constrictions. Concentration and trapping of DNA was observed to depend on the properties of the suspending medium and the frequency of the applied electric field.[10]

In this work, we present the manipulation and concentration of polystyrene microparticles with an AC electric in a microchannel containing an array of cylindrical posts along a microchannel. The results demonstrated that the electric field frequencies needed for manipulation are significantly smaller when compared to results obtained in other studied reported in the literature.[4]

## Dielectrophoresis Theory

Consider a spherical particle of radius $r$ suspended in a dielectric medium with permittivity $\varepsilon_m$. In the presence of non-uniform electric field with magnitude $E$, the particle is subject to a force due to DEP, described by:

$$F_{DEP} = 2\pi r^3 \varepsilon_m \operatorname{Re}[f_{CM}]\nabla E^2 \quad (1)$$

where, $f_{CM}$ is de Clausius-Mossotti (CM) factor. For an alternating electric field with frequency $\omega$, the CM factor is related to the complex permittivity ($\varepsilon^* = \varepsilon + \sigma/i\omega$) of both the particle and the suspending medium as:

$$f_{CM} = \frac{\varepsilon_p^* - \varepsilon_m^*}{\varepsilon_p^* + 2\varepsilon_m^*} \quad (2)$$

If diffusion and pressure driven flow are neglible, the motion of a particle subject to a non-uniform electric field can be described by the electrokinetic (EK) flow and the flow due to DEP. The EK velocity is proportional to the applied electric field as:

$$\vec{v}_{EK} = \mu_{EK}\vec{E} \quad (3)$$

where, $\mu_{EK}$ is the electrokinetic mobility, and it takes account for the effects of electroosmosis and electrophoresis. For a particle with small surface charge, the effect of electrophoresis can be neglected. Under this assumption, the electrokinetic mobility depends only on the properties of the channel and the suspending medium. The velocity of the particle due to DEP can be described as:

$$\vec{v}_{DEP} = -\mu_{DEP}\nabla E^2 \quad (4)$$

Trapping of particles is achieved when the effects of DEP on particle flow overcomes the EK contribution. This condition can be expressed in the following relation:

$$\frac{\mu_{DEP}\nabla E^2}{\mu_{EK}E^2} \cdot \vec{E} > 1 \quad (5)$$

## Materials and Methods

### Mathematical Model

Figure 1 presents the microchannel considered for this study. The geometry of a microfluidic channel was captured into *COMSOL Multiphysics* software. The micro channel considered was 10.12-mm long, and 2-mm wide, containing an array of 32 insulating posts, distributed in 4 rows of 8 posts each. The posts had a diameter of 440 μm, arranged 520 μm center-to-center (Figure 1).Particles with a diameter of 1 μm, and conductivity of 10 $\mu Scm^{-1}$ were considered. Also no surface charge was taken into account in the model. A triangular mesh was built in order to solve the Laplace equation using the *Lagrange element* implemented in the software. From the obtained numerical solution it is possible to obtain values for the electrical field and the gradient of $E^2$.

### Experimental

The microfluidic channel employed for experimentation was fabricated from Schott D263 glass wafers, and had the geometry described above and it is shown in Figure 1. The microparticles employed were carboxylated-modified fluorescent green-yellow 1-μm-diameter microspheres

(Invitrogen, Carlsbad, CA, USA), with a concentration of 7.639 x $10^{10}$ microspheresml$^{-1}$. These particles contained a fluorescent dye with excitation/emission wavelengths of 505/515 nm. Image and video capturing was done employing an inverted microfluidic microscope SVM340 with an integrated video camera (LabSmith, Livermore, CA, USA). A high voltage power supply HVS448 (LabSmith, Livermore, CA, USA) was also used to apply AC electric fields employing platinum-wire electrodes. The amplitude, period and time variation of the alternating field was adjusted and controlled employing the software of the power supply.

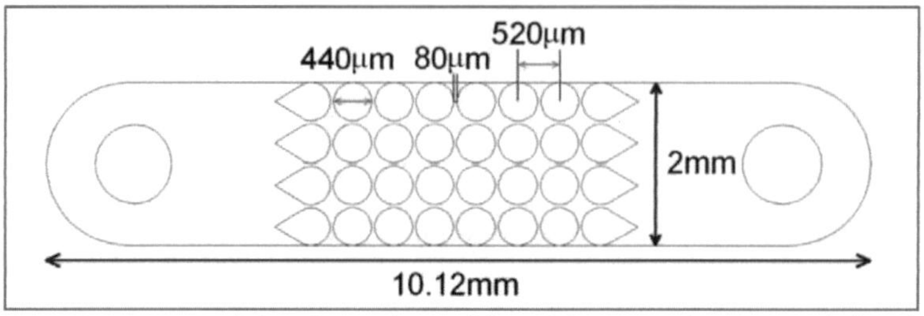


**Figure 1.** Geometry of the channel considered for experiments and simulation.

## Results and Discussion

### Simulation

Figure 2 shows the magnitude of the time-averaged $\nabla E^2$ and its x-component between the insulating cylindrical posts, where the highest magnitude is reached at the narrow region between posts. It is interesting to note that, for a sinusoidal time dependence, the time averaged magnitude of $\nabla E^2$ takes values always greater than zero, as shown in Figure 2a. On the other hand, Figure 2b shows how the x-component can take positive and negative values, although this variation is only spatially dependant. In contrast, while the time average of $E$ is zero, its x component, can take positive or negative values, depending on time. Thus, from the relation described in (5) it is expected that the regions where particles are concentrated due to DEP change their location due to its time dependence.

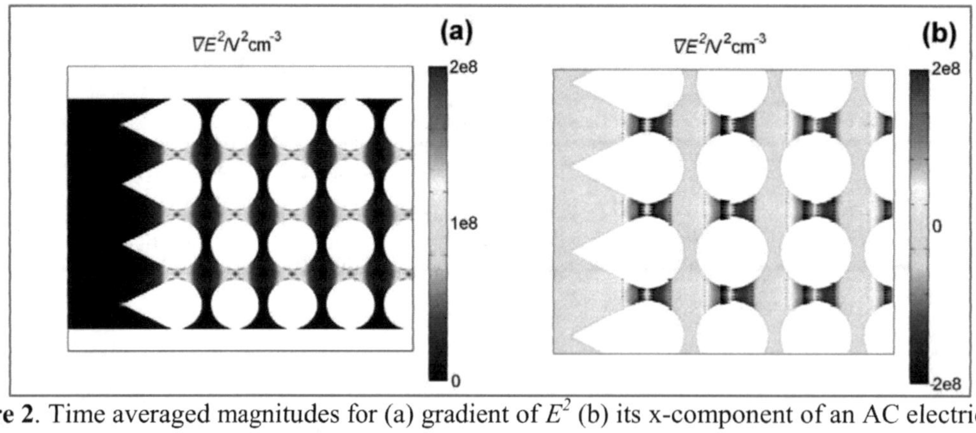


**Figure 2**. Time averaged magnitudes for (a) gradient of $E^2$ (b) its x-component of an AC electric field with amplitude of 750 V and a period of 1000 ms (1 Hz).

**Experimental procedure**

A series of experiments were carried out combining DC and AC electric fields. A DC field was employed to introduce microparticles into the microchannel with an electroosmotic flow and to induce DEP trapping of particles, with an applied potential of 750 V. After 10 seconds, the signal was switched to AC with 750 V of amplitude and different periods. The time variation had a sinusoidal dependence. Figure 3 shows the profile of the band of particles trapped by DEP for different times when the period of the applied alternating electric potential is 1000 ms (1 Hz). It is observed that the band of trapped particles (Fig 3a) has a periodic behavior: the band splits (Fig 3b) and then is reformed between adjacent posts (Figs. 3c and 3d), and so forth. Particles seem to keep bouncing from one trapping region to the next, moving a distance that is approximately equal to the center-to-center separation of the posts. It is observed that particles occupy the corresponding regions of trapping, according to the direction of the electrokinetic flow.

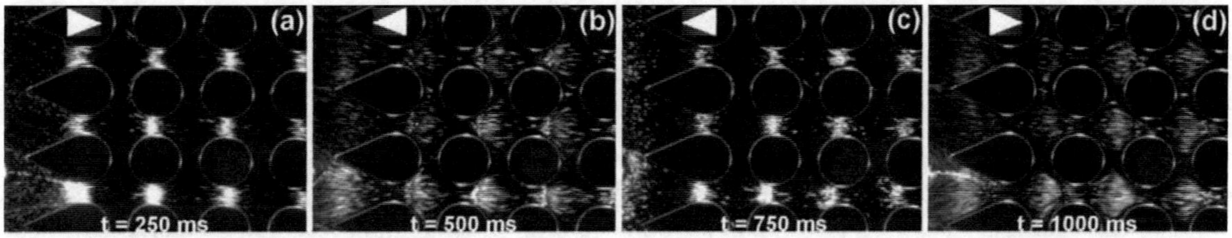


**Figure 3**. Manipulation of trapping regions between posts when a sinusoidal signal with period T=1000 ms (1 Hz) is applied. Arrow indicates direction of flow. (a) Particles being trapped (b) particles moving from right to left (c) band of trapped particles generated again and, (d) particles moving in the positive direction.

The periodic behavior is more clearly observed in Figure 4, where the applied electric potential has a frequency of 2500 ms (0.4 Hz). Here, after forming bands (Fig 4a), particles are able to move (Fig 4b) a distance that is greater than the center-to-center separation of adjacent posts. In this case, the bands of trapped particles are reformed (Fig 4c) in a more clearly manner, just before particles are free to move again (Fig 4d).

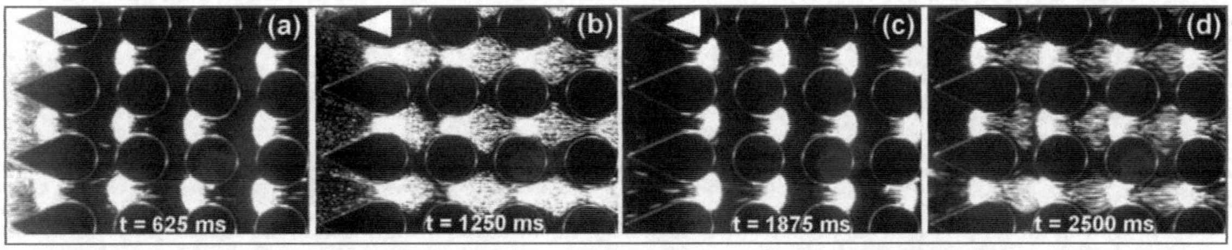


**Figure 3**. Manipulation of trapping regions between posts when a sinusoidal signal with period T=2500 ms (0.4 Hz) is applied. Arrow indicates direction of flow. (a) Bands of trapped particles in the positive flow (b) particles moving in the negative direction (c) bands of trapped particles are generated (d) particles moving in the positive direction, before forming trapping bands again.

## Conclusions

In this work, it is demonstrated the manipulation and concentration of polystyrene microparticles employing an AC electric potential in a microchannel with an array of cylindrical insulating structures. The location of bands of trapped particles was studied for different frequencies of the applied electric potential. The frequencies required to achieve manipulation are small when compared to frequencies reported in other studies that employ AC electric fields. This work demonstrates that it is possible to combine AC electric fields and insulating structures for manipulation of particles successfully.

## Acknowledgments

The authors wish to thank the financial support given by the grant CONACYT-CB-2006-53603, and *Cátedra de Investigación* of *Tecnológico de Monterrey* (CAT 142).

## References

1 Pohl H.A., (1951), "The Motion and Precipitation of Suspensoids in Divergent Electric Fields," *J. Appl. Phys.* 22(7), pp.869-871.

2 Lapizco-Encinas B.H., Rito-Palomares M., (2007), "Dielectrophoresis for the manipulation of nanobioparticles," *Electrophoresis* 28(24), pp.4521-4538.

3 Voldman J., (2006), "Electrical forces for microscale cell manipulation," *Ann. Rev. Biomed. Eng.* 8, pp.425-454.

4 Gascoyne P.R.C., Vykoukal J., (2002), "Particle separation by dielectrophoresis," *Electrophoresis* 23(13), pp.1973-1983.

5 Gray D.S., Tan J.L., Voldman J., Chen C.S., (2004), "Dielectrophoretic registration of living cells to a microelectrode array," *Biosens. Bioelectron.* 19(7), pp.771-780.

6 Abidin Z.Z., Markx G.H., (2005), "High-gradient electric field system for the dielectrophoretic separation of cells," *J. Electrostatics* 63(6-10), pp.823-830.

7 Lapizco-Encinas B.H., Simmons B.A., Cummings E.B., Fintschenko Y., (2004), "Dielectrophoretic Concentration and Separation of Live and Dead Bacteria in an Array of Insulators," *Anal. Chem.* 76(6), pp.1571-1579.

8 Lapizco-Encinas B.H., Simmons B.A., Cummings E.B., Fintschenko Y., (2004), "Insulator-Based Dielectrophoresis for the Selective Concentration and Separation of Live Bacteria in Water," *Electrophoresis* 25, pp.1695-1704.

9 Cummings E.B., "A Comparison of Theoretical and Experimental Electrokinetic and Dielectrophoretic Flow Fields," (2002), in 32nd AIAA Fluid Dynamics Conference and Exhibit 2002, American Institute of Aeronautics and Astronautics, St. Louis, Missouri, vol. paper 2002-3193,, pp. 1-17.

10 Chou C.F., Tegenfeldt J.O., Bakajin O., Chan S.S., Cox E.C., Darnton N., Duke T., Austin R.H., (2002), "Electrodeless dielectrophoresis of single- and double-stranded DNA," *Biophys. J.* 83(4), pp.2170-2179.

# Capillary Electrophoresis as a Tool to Monitor the Early Stages of Insulin Aggregation

**Elizabeth Pryor**, Chemical Engineering, University of Arkansas, Fayetteville, AR

Christa Hestekin, Chemical Engineering, University of Arkansas, Fayetteville, AR

Melissa A. Moss, Chemical Engineering, University of South Carolina, Columbia, SC

Protein folding abnormalities contribute to many amyloid diseases such as: Alzheimer's disease, Parkinson's disease, Type II diabetes mellitus, prion diseases, and Huntington's disease. Insoluble protein aggregates, termed amyloids, which exhibit a cross beta-sheet structure are a common characteristic of amyloid diseases. The molecular mechanism by which proteins progress from monomer to oligomer to insoluble amyloid aggregates is not well understood. The formation of low molecular weight oligomeric species during the early stages of protein misfolding is of particular interest for disease prevention. Currently, diagnostic techniques which monitor the very early stages of protein misfolding remain inadequate. Insulin protein is known to form amyloid aggregates displaying a cross beta-sheet structure. In this work, capillary electrophoresis is used to monitor the progression of insulin protein from monomeric to low molecular weight oligomeric species. A critical aggregation time period of 8 to 10 hours was found to be necessary for the formation of low molecular weight oligomeric species, however this time period was dependent on aggregation conditions such as the amount of salt. The limit of detection of insulin using capillary electrophoresis detection by both UV and laser induced fluorescence was determined. By determining the early aggregation kinetics of insulin at physiologically relevant concentrations (nM-pM range), we have shown that capillary electrophoresis has the ability to detect other disease causing amyloid proteins, similar to insulin, present in cerebrospinal fluid or plasma.

See more of this Session: Poster Session: Annual Meeting of the American Electrophoresis Society
See more of this Group/Topical: Topical 3: 2009 Annual Meeting of the American Electrophoresis Society (AES)

# Nanocomposite Gels for Improved Separations in Clinical Diagnostics

**Jeffery W. Thompson**, Chemical Engineering, Tennessee Technological University, Cookeville, TN

Holly Stretz, Chemical Engineering, Tennessee Technological University, Cookeville, TN

Pedro Arce, Chemical Engineering, Tennessee Technological University, Cookeville, TN

Gels are widely used in biological separations, but commonly certain protein or oligosaccharide mixtures may exhibit low resolution. Separation control mechanisms include sieving (morphological control) or electrosmosis for electrophoretic separations. Charged gold nanoparticles of varying aspect ratios, measured by transmission electron microscopy, have been embedded in poly-acrylamide gels. Proteins such as ovalbumin and BSA show unique mobilities and separations in composite gels versus control gels. The effets of aspect ratio on protein mobilities will be discussed.

See more of this Session: Poster Session: Annual Meeting of the American Electrophoresis Society
See more of this Group/Topical: Topical 3: 2009 Annual Meeting of the American Electrophoresis Society (AES)

# Induced-Charge Electrophoresis of Metallodielectric Janus Particles

**Sumit Gangwal**, Department of Chemical and Biomolecular Engineering, North Carolina State University, Raleigh, NC

Olivier J. Cayre, Department of Chemical and Biomolecular Engineering, North Carolina State University, Raleigh, NC

Martin Z. Bazant, Chemical Engineering and Mathematics, Massachusetts Institute of Technology, Cambridge, MA

Orlin D. Velev, Department of Chemical and Biomolecular Engineering, North Carolina State University, Raleigh, NC

The synthesis of "Janus" particles (whose hemispheres are physically or chemically different) is of growing importance for the development of novel materials, but the behavior of such particles in external fields has not been studied in depth. The application of AC electric fields in aqueous suspensions of anisotropic particles leads to unbalanced liquid flows and nonlinear, induced-charge electrophoretic (ICEP) motion. We report experimental observations of the motion of "Janus" microparticles with one dielectric and one metal-coated hemisphere induced by uniform fields of frequency 100 Hz - 10 kHz in NaCl solutions. The motion is perpendicular to the field axis and persists after particles are attracted to a glass wall. The velocity as a function of field strength, AC frequency and particle size are consistent with ICEP theory in dilute solutions ($\leq 100\ \mu$ M), but we observe anomalous behavior at higher concentrations, including no motion above 10 mM. The metallodielectric Janus particles' motion across uniform AC electric field lines is, to our knowledge, the first report of this fascinating phenomenon that can now guide further development of nonlinear electrokinetic theory. The propelling metallodielectric particles may find applications as microscopic mixers, "shuttles", self-propelling on-chip sensors, microactuators and microsensors in MEMS devices, optoelectronic devices or for separations in microfluidic devices.

See more of this Session: Poster Session: Annual Meeting of the American Electrophoresis Society
See more of this Group/Topical: Topical 3: 2009 Annual Meeting of the American Electrophoresis Society (AES)

# Effect of Material Morphology On Optimal Time of Separation of Bio-Molecules

**Jyothirmai J. Simhadri**, Chemical Engineering, Tennessee Technological University, Cookeville, TN

Mario Oyanader, Chemical Engineering, Tennessee Technological University, Cookeville, TN

Holly Stretz, Chemical Engineering, Tennessee Technological University, Cookeville, TN

Pedro Arce, Chemical Engineering, Tennessee Technological University, Cookeville, TN

Electrophoresis in polymer hydrogels with nanometer-scale pore structure are widely used for the separation and purification of biological macromolecules. In gel-electrophoresis, the internal morphology of the gel also plays an important role in improving the separation. Tuning the nanometer-scale pore structure of the gel either by templating or by adding nanoparticles to improve separations has been the current area of focus in our research group. Moreover, analysis of the effects of the nature of the pore alignment, pore length and diameters on the transport of macromolecules is an important aspect to be studied either analytically or computationally as shown by previous efforts (Trinh et al, 1999; Hidalgo et. al, 2007). In this research, non-uniform channel models whose characteristics are useful domains to mimic the gel morphology are analyzed computationally. The results illustrate the effects of the various geometrical parameters with electric fields on effective transport parameters and eventually determine the optimal time of separation.

References:

1. Trinh, S., B.R. Locke and P.E. Arce, "Convective and Electroconvective Transport in Non-Uniform Channels with Application to Macromolecular Separations," Separation and Purification Technology, 15, 255 (1999).

2. Hidalgo, R., M. A. Oyanader (*), and P. E. Arce, "Dispersive Mixing Effect Caused by Combined Effect of Channel Morphology and Electrophoretic Mobility in Poiseuille Flows." AIChE Annual Meeting, Salt Lake City, 2007. (*) Speaker.

See more of this Session: Poster Session: Annual Meeting of the American Electrophoresis Society
See more of this Group/Topical: Topical 3: 2009 Annual Meeting of the American Electrophoresis Society (AES)

# Comparison of High Abundance Protein Depletion Techniques

**Aran Paulus**, Bio-Rad Laboratories, Hercules, CA

Tim Wehr, Bio-Rad Laboratories, Hercules, CA

Chengjun Sun, Bio-Rad Laboratories, Hercules, CA

Lei Li, Bio-Rad Laboratories, Hercules, CA

Katrina Academia, Bio-Rad Laboratories, Hercules, CA

Steve Freeby, Bio-Rad Laboratories, Hercules, CA

Ning Liu, Bio-Rad Laboratories, Hercules, CA

John Walker, Bio-Rad Laboratories, Hercules, CA

Chris Sutton, Institute of Cancer Therapeutics,, University of Bradford, Bradford, Bradford, United Kingdom

To identify candidate biomarkers present in human plasma at moderate to low abundance, high abundance proteins (HAPs) must first be removed. Several methods have been described for HAP removal, including antibody affinity, peptide affinity, and ion exchange. The goal of this study was to compare one popular antibody-based affinity method with a newer peptide-based affinity method for HAP removal from human plasma. Two proteomic workflows were used. The first employed separation HAP-depleted eluates on a one-dimensional PAGE gel, followed by in-gel digestion of lane segments and reversed-phase phase LC-ESI-MS-MS ("GeLC-MS"). The second workflow employed tryptic digestion of HAP-depleted eluates followed by nano-reversed phase HPLC and MALDI-TOF-TOF analysis of chromatographic fractions. This study is the first direct comparison of an antibody affinity-based depletion method and a peptide affinity-based depletion method looking at the entire proteomics workflow as used in biomarker discovery programs

See more of this Session: Advances in Proteomics: New Technologies
See more of this Group/Topical: Topical 3: 2009 Annual Meeting of the American Electrophoresis Society (AES)

# Fluorescent Zdye Platform for Differential Covalent Labeling of Proteins for 2D and 1D Gel Analysis

**Edward Dratz**, Chemistry and Biochemistry, Montana State University, Bozeman, MT

Paul Grieco, Chemistry and Biochemistry, Montana State University, Bozeman, MT

Mary Cloninger, Chemistry and Biochemistry, Montana State University, Bozeman, MT

Kristian Schlick, Chemistry and Biochemistry, Montana State University, Bozeman, MT

Duane Mooney, Chemistry and Biochemistry, Montana State University, Bozeman, MT

Ting Liu, Chemistry and Biochemistry, Montana State University, Bozeman, MT

Matt Shipman, Chemistry and Biochemistry, Montana State University, Bozeman, MT

Scott Laffoon, Chemistry and Biochemistry, Montana State University, Bozeman, MT

Jared Bowden, Chemistry and Biochemistry, Montana State University, Bozeman, MT

Ben Reeves, Chemistry and Biochemistry, Montana State University, Bozeman, MT

Kevin Spicka, Chemistry and Biochemistry, Montana State University, Bozeman, MT

Hannes Helmboldt, Chemistry and Biochemistry, Montana State University, Bozeman, MT

Ben Skerik-Borg, Chemistry and Biochemistry, Montana State University, Bozeman, MT

Brian Richards, Chemistry and Biochemistry, Montana State University, Bozeman, MT

Casey Smith, Resonon, Inc., Bozeman, MT

Rand Swanson, Resonon, Inc., Bozeman, MT

A new array of fluorescent dyes (Zdyes) has been developed for differential, covalent labeling of proteins in 2D and 1D gel analysis. Zdyes contain zwitterionic charge pairs, which increase the solubility of labeled proteins at the protein's isoelectric points. The absorbance and fluorescence spectra of the Zdyes allow the use of widely available laser-excited fluorescence scanners.

The amine reactive Zdyes contain a titrable tertiary amine whose pK approximates that of the epsilon amine of lysine residues, preserving the pI of the unlabeled proteins. Zdyes containing a maleimide or a pyridine disulfide moiety react with protein thiols and can be used for saturation labeling, while increasing protein solubility for more sensitive 2D gel analysis. We are also developing Zdyes and associated chemistries to covalently bind multicolor Zdyes to phosphorylated, glycosidated, and nitrosylated sites in proteins for differential detection. Multicolor Zdyes that bind to active sites of native proteins, such as proteases and protein phosphatases, are also being developed.

The combination of high quantum yields and extinction coefficients of Zdyes with their increased solubility allows a deeper view into the proteome via 2D gels than is available with alternative fluorescent dyes. In 1D gels amine-labeled protein bands containing as little as 20pg of protein can be measured with confidence. Zdyes have been implemented in several projects. Using 2D gels the serum proteome of Type II diabetic subjects relative to control groups has been probed for differences in protein levels and post-translational modifications. The effects of infection of cultured monocytes by the bacterium Coxiella burnetii are also being studied using Zdyes. Differential protein expression in epileptic and normal human brain is also being examined. Complex protein mixtures benefit from the use of long, serial IEF to increase 2D resolution and spot count especially when the amount of protein is limiting. A new hyperspectral gel scanner allows the measurement of full fluorescent spectra for each image pixel, and provides accurate deconvolution of overlapping fluorescent dye emission spectra, accurate fluorescent background for each pixel, increased sensitivity and dynamic range.

Taken together this new Zdye platform allows highly sensitive fluorescent staining of proteins based on covalent binding to amines and thiols, and to sites of posttranslational modifications, making this platform compatible with a wide range of experimental applications.

See more of this Session: Advances in Proteomics: New Technologies
See more of this Group/Topical: Topical 3: 2009 Annual Meeting of the American Electrophoresis Society (AES)

# Providing "Freeze Frames" of Cellular Processes by Thermal Stabilization of Samples: The Inactivation of Proteases, Preservation of Molecular Complexes, and Elimination of Apocryphal Spots on Two-Dimensional Gels

*Gary B. Smejkal, Harvard Clinical and Translational Science Center, Boston, MA.*

## Abstract

An understanding of cellular processes is facilitated only when methods such as two-dimensional gel electrophoresis (2DGE) are capable of isolating without bias the entire protein constituency of cells such that the analysis can provide a comprehensive representation of the proteome. The regulated expression of proteins as it relates to normal or disease processes, the efficacy of therapeutic regime, or the identification of phenotypes requires that specific proteins are accurately quantified. Therefore, reliable methods of sample preparation become necessary to produce meaningful results. Since proteases regulate many cellular processes including the clearance of down-regulated proteins and the cleavage of precursors to produce functional proteins, it is essential to completely abolish all enzymatic activity to elicit a realistic "snapshot" of the proteome as it occurs *in vivo*, both qualitatively and quantitatively. Further, the phosphorylation states of proteins may be altered if enzymes involved in phosphorylation and dephosphorylation are not promptly inactivated. Typically, proteolytic activity is curtailed by rapidly freezing the sample and by the inclusion of protease inhibitors and metal chelators. However, freezing preserves the activity of enzymes which is restored when the samples are thawed. The formation of apocryphal protein spots in 2D gels indicates the inability to completely halt enzyme activity before protein fragmentation can occur. Further, alterations of the proteome caused by *in vitro* degradation may obscure important information of a sample's biological state and hinder the discovery of biomarkers. An alternative is to use rapid heat inactivation to eliminate enzymatic activity. The Denator Stabilizor T1 has been evaluated and shown to improve results by decreasing proteolytic fragments.

## Introduction

Protein degradation in biological samples confounds proteomic analyses. Rapid thermal inactivation of intrinsic enzymatic activity has been evaluated using murine brain and 2DGE. It is found to be fully compatible with the downstream analysis as well as limiting proteolytic fragmentation and preserving phosphorylation states of some proteins. Modifications of the proteome constituency such as degradation obscures important information of a sample's biological state, hinders the discovery of biomarkers, and diminishes accuracy when trying to determine relative differences in protein expression. In 2DGE, chemical inhibitors are routinely added to urea and detergent based reagents used for protein extraction to ensure minimal protein degradation. An alternative is to use thermal inactivation to eliminate enzymatic activity. In these studies, thermal inactivation has been is shown to minimize proteolysis as well as to prevent the observed pI shifts of phosphoprotein trains due to residual phosphomonoesterase or kinase activity.

## Methods

### Stabilization of samples

Mice brains (*n*=6) were extracted after cervical dislocation, dissected into halves, and each hemisphere was stabilized within two minutes by either (i) heat inactivation in the Denator Stabilizor T1, (ii) freezing in liquid N2, or (iii) chemical inactivation using broad range protease inhibitors. All samples were processed within two minutes of dissection.

### Solubilization of proteins for 2DGE

Proteins from the preserved tissues were extracted in 7M urea, 2M thiourea, 4% CHAPS, and 5 mM TBP using a Pressure Cycling Technology [1,2] from Pressure Biosciences (South Easton, MA, USA). Insoluble material was pelleted centrifugally and the supernatants were alkylated two hours with 10 mM acrylamide monomer. The alkylation reaction quenched by the addition of 50 mM DTT. IPG strips pH 3-10 were hydrated with 200 uL of each protein solution normalized to 1 mg/mL, followed by IEF and 2DGE as previously described by Smejkal *et al.* [3].

### Phosphoprotein staining and image analysis

Gels were stained with Pro-Q Diamond Fluorescent Phosphoprotein Stain (Invitrogen, Carlsbad, CA, USA), followed by secondary staining for total protein using SYPRO Ruby or Flamingo stain (BioRad, Hercules, CA, USA). Images were captured using the BioRad VersaDoc Fluorescent Imager and analyzed using the Ludesi REDFIN SOLO image analysis software (Ludesi Corporation, Lund, Sweden). Quantitation of protein spots was derived from duplicate gels ($r^2$ values ranged from 0.9925 to 0.9962 with spots having integrated volumes less than 300 filtered from the analyses) and was expressed as mean integrated spot fluorescence.

## Results and Discussion

Previously, Grinyer *et al.* [4] showed that residual protease activity in samples prepared for 2DGE resulted in the formation of artifactual spots in 2D gels. Smejkal *et al.* [2] showed that metal ions released from sonicator probes reactivate some proteases, even in the presence of protease inhibitors. Likewise, clot formation is frequently observed in plasma samples diluted in urea, thiourea, and CHAPS where thrombin activity is expected to be abolished. The need to completely inactivate enzymes prior to 2D analyses is exemplified by the microcrustacean *Daphnia magna*, where proteolysis is so rapid and extensive that live organisms must be steeped in excess protease inhibitors [5].

More recently, Robinson *et al.* [6] demonstrated that thermal inactivation of proteases corresponds with the preservation of high molecular weight proteins and a reduction in protein fragmentation, corroborated by the identification of protein fragments by mass spectrometry.

Svensson *et al.* [7] showed levels of phosphorylated CREB, GSK and MAPK in mouse brain cortex were preserved following thermal inactivation in the Stabilizor T1, whereas the levels of these proteins were significantly diminished in unstabilized tissues. In these studies, we show numerous instances of shifts in protein molecular weight and pI that are apparently linked to incomplete inactivation of proteases and phosphatases, respectively. Mass spectrometry is ongoing for the identification of specific protein spots exhibiting pI shifts and the analysis of their phosphorylation state.

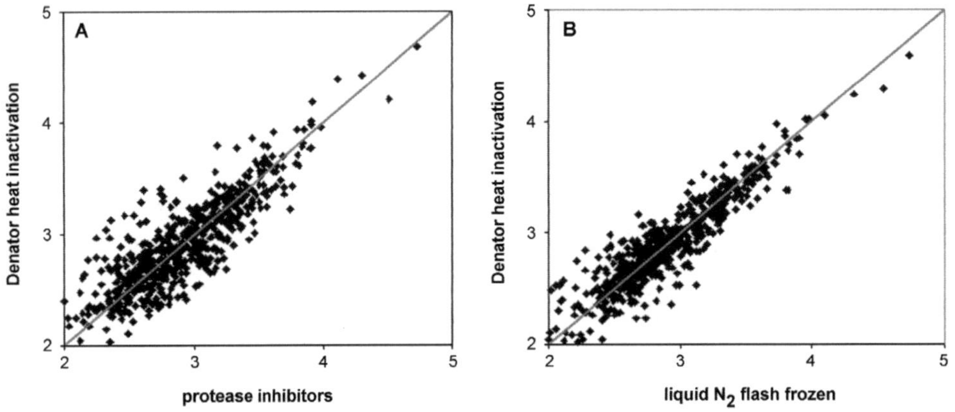


**Figure 1.** Protein staining (log mean integrated spot fluorescence) of 612 matched protein spots in 2D gels comparing tissue preservation methods. (A) heat inactivation vs. chemical protease inhibition, or (B) heat inactivation vs. flash freezing. Shifts from the theoretical slope of 1.00 (red line) demonstrate changes in spot intensity in heat inactivated samples.

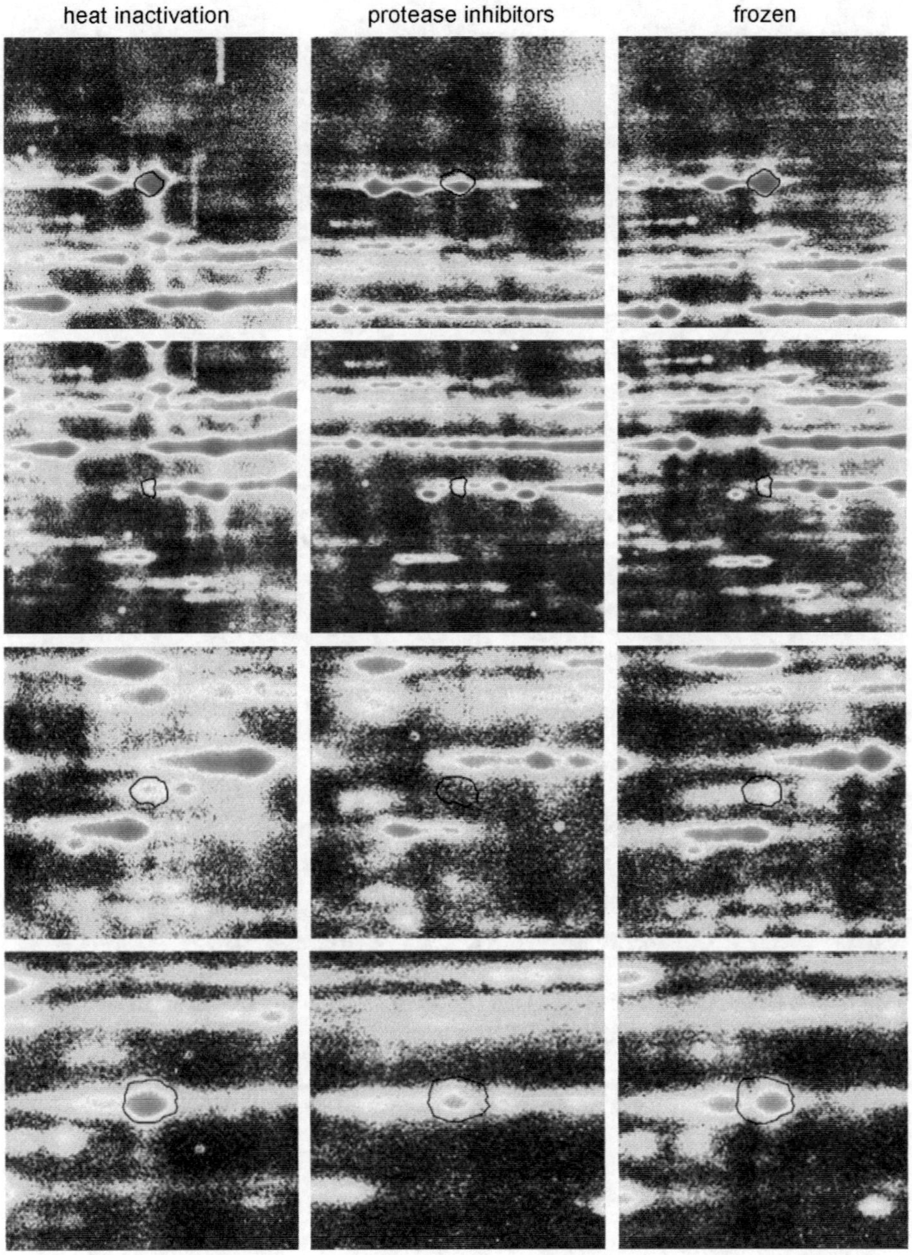


**Figure 2.** Image analysis of enlarged regions of 2D gels showing shifts in protein pI in samples preserved by heat activation (left), chemical inactivation with protease inhibitors (center), or flash freezing in liquid nitrogen (right). Proposed shifts in pI are due to incomplete inhibition of kinases or phosphatases.

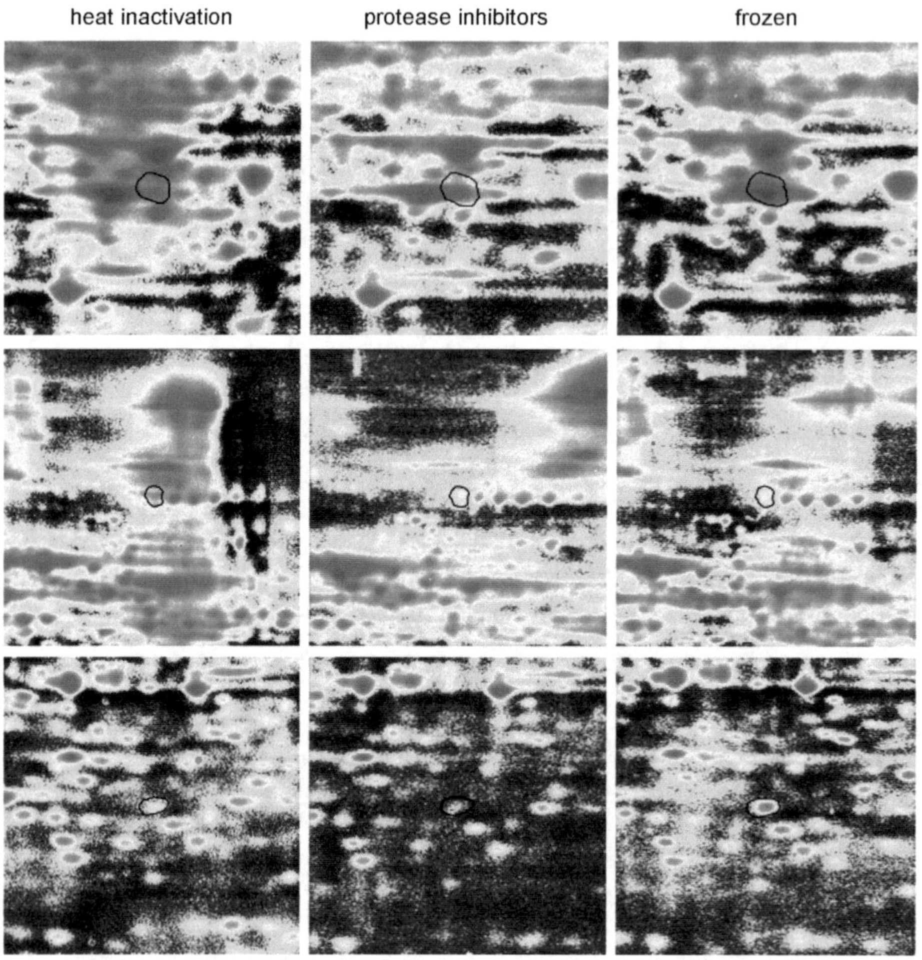


**Figure 3.** Image analysis of enlarged regions of 2D gels showing shifts in protein molecular weights in samples preserved by heat activation (left), chemical inactivation with protease inhibitors (center), or flash freezing in liquid nitrogen (right). Proposed shifts in molecular weight are due to incomplete inhibition of proteases.

## References

[1] Ringham *et al.* (2007). *Electrophoresis* 2007, 28, 1022-1024.
[2] Smejkal *et al.* (2007). *Anal. Biochem.* 2007, 363, 309-311.
[3] Smejkal *et al.* (2006). *J. Proteome Res.* 2006, 5, 983- 987.
[4] Grinyer *et al.* (2004). *Curr. Genet.* 45, 163-169.
[5] Bauer *et al.* (2009). *HUPO VIII World Congress,* Toronto Canada.
[6] Robinson *et al.* (2009). *Proteomics*, published online.
[7] Svensson *et al.* (2009). *J. Proteome Research* 8, 974-981.

# Optimization of 2D Gel Transblotting of Complex Protein Mixtures

**Nancy Kendrick**, Kendrick Labs Inc, Madison, WI

Jon J. Johansen, Kendrick Labs Inc, Madison, WI

Matthew Hoelter, Kendrick Labs Inc, Madison, WI

2D gel Western blotting of Host Cell Proteins (HCP) is becoming an important method for characterization of recombinant protein products. However the percentage of protein transferred from a 1D or 2D polyacrylamide gel to PVDF or nitrocellulose membrane depends strongly on conditions of transfer. While optimization of protein transblotting conditions for a few standard proteins is sufficient for many research problems, for HCP samples it is important to maximize recoveries for proteins of varied molecular weights and isoelectric points in complex mixtures. We will present optimization results for 2D gel patterns of protein mixtures based on Coomassie blue staining of PVDF membranes. In addition we will show optimization results based on HCP Western blotting as an assay. Result s for varied buffer pH and transblotting time will be shown.

See more of this Session: Advances in Proteomics: New Technologies
See more of this Group/Topical: Topical 3: 2009 Annual Meeting of the American Electrophoresis Society (AES)

# Acid-Degradable Polyacrylamide Gel Electrophoresis for Isolation of Structurally and Functionally Intact Proteins and Its Implications

Yoon Kyung Kim, Biomedical Engineering, University of California, Irvine, Irvine, CA

**Young Jik Kwon**, Pharmaceutical Sciences, Chemical Eng & Materials Science, Biomedical Eng, University of California, Irvine, Irvine, CA

In post-genomic era, understanding protein structure/function coupled with genomics, called proteomics, is an emerging technology not only in understanding biology at a molecular level but also in discovering biomarkers for diagnosis and therapy. Pivotal steps in current proteomic studies are to separate proteins from a highly complex biological sample and to analyze the isolated proteins using various analysis techniques such as MALDI (matrix-assisted laser desorption/ionization). Polyacrylamide gel electrophoresis (PAGE), conventionally applied with or without sodium dodecyl sulfate (SDS) in denatured or native PAGE, respectively, has been the predominant technique in simultaneously separating a number of proteins. However, significantly larger size of proteins than the pores in a polyacrylamide gel makes recovering the separated proteins in their structurally and functionally native forms extremely inefficient or almost impossible, particularly for large proteins. Frequently, the proteins are fragmented by proteolysis in order to be released by diffusion (in-gel digestion), resulting in obstructed elucidation of native structure-function relationships of a separated protein. In this study, we tested the novel idea of obtaining structurally and functionally intact proteins by degrading polyacrylamide gel (rather than fragmenting separated proteins) under a mildly acidic condition. A model large protein, mouse IgG1 antibody (MW 150 kDa) was electrophoresed in the polyacrylamide gel cross-linked by acid-cleavable ketal diacrylamide instead of conventionally used methylene bisacrylamide, under native PAGE conditions. A protein band identified at 150 kDa was excised and hydrolyzed at the acidic solution consisting of formic acid and propanol in water. The proteins were further purified by selectively precipitating hydrolyzed polyacrylamide. The final amount of the isolated proteins quantified by enzyme-linked immunosorbent assay (ELISA) demonstrated that approximately 70% of the antibodies were recovered from the degradable polyacrylamide gel, while only a traceable amount of the antibodies were obtained from a conventionally used nondegradable polyacrylamide gel. The preserved specific binding capability of the isolated antibodies to target antigens as well as intact molecular weights proved the successful isolation of structurally and functionally native proteins. This novel technology can be used to isolate other types of macromolecules, macromolecular complexes, and nanomaterials.

See more of this Session: Advances in Proteomics: New Technologies
See more of this Group/Topical: Topical 3: 2009 Annual Meeting of the American Electrophoresis Society (AES)

# A De Novo Approach for Untargeted Post-Translational Modification Prediction Using Tandem Mass Spectrometry and Integer Linear Optimization

**Richard Baliban**, Chemical Engineering, Princeton University, Princeton, NJ

Peter A. DiMaggio Jr., Chemical Engineering, Princeton University, Princeton, NJ

Nicolas Young, Molecular Biology, Princeton University, Princeton, NJ

Benjamin A. Garcia, Molecular Biology, Princeton University, Princeton, NJ

Christodoulos A. Floudas, Chemical Engineering, Princeton University, Princeton, NJ

Classification of the post-translational modifications (PTMs) of various organisms is currently a major challenge in the field of proteomics. Tandem mass spectrometry (MS/MS) has shown to be an excellent tool for peptide sequence prediction, and is indispensable in determining PTMs [1,2]. Consequentially, many research groups [3-9] have incorporated modification discovery into their respective identification algorithms which utilize multiple databases to build a list of variable modifications that can exist on a candidate peptide. To circumvent the large complexity associated with variable modification identification, an upper bound is often assigned for the total number of (a) variable modification types or (b) variable modification sites. However, thorough determination of a dynamic proteome requires an unbiased analysis of any and all modifications that can occur. Unrestrictive algorithms have been developed [3,4] which focus on the spectral alignment to create integer mass modifications that exist on a candidate peptide. However, these algorithms place a restriction on the number of modification sites to enhance computational efficiency and reduce the false detection of low-mass modifications.

We have developed PILOT_PTM, which is a novel de novo method for untargeted PTM Prediction via Integer Linear Optimization (ILP) and Tandem mass spectrometry. ILP has been an integral tool in the highly accurate de novo sequencing algorithm PILOT [10,11] and the hybrid algorithm PILOT_SEQUEL [12]. Similar to these methods, our objective function seeks to maximize the sum of the intensity contributions from theoretical peak matches to the experimental spectrum given logical constraints. Given a template amino acid sequence, the method can predict the modifications at all positions on the peptide sequence. That is, our model will assume that all template positions can be post-translationally modified and will seek to determine the optimal set of modifications among a universal list based on the MS/MS data. PILOT_PTM is not limited to choosing modification sites for a peptide and will not infer a sequence tag from the MS/MS. The method rigorously guarantees the optimal set of modifications without having to enumerate all combinations of possible modifications.

To verify the capability of the method across distinct data types, PILOT_PTM was tested on several data sets including (a) unmodified peptides fragmented via Collision Induced Dissociation (CID) on (i) ion trap, (ii) Q-TOF, and (iii) Orbitrap mass spectrometers, (b) chemically synthesized phosphopeptides fragmented via Electron Transfer Dissociation (ETD), (c) the 1-50 N-terminal tail of the histone H3 protein fragmented via Electron Collision Dissociation (ECD), and (d) propionylated H3 peptides fragmented via CID. The results of each data set were quantified using several metrics including the overall residue prediction accuracy, the complete peptide prediction accuracy, and the subsequence accuracy. PILOT_PTM maintains a high accuracy for all scoring metrics across all data sets, emphasizing the capability as an instrument independent algorithm capable of handling multiple fragmentation methods.

To benchmark the capability of the method, the results of the (d) propionylated H3 data set were compared with those acquired from five state-of-the-art algorithms including three hybrid sequence tag/database approaches (InsPecT [5], Modi [6], and VEMS [7]) and two pure database approaches (Mascot [8] and X!Tandem [9]). To develop a method for comparison, a protocol was established for Mascot, InsPecT, VEMS, and X!Tandem. Since these algorithms all severely restrict the number of variable modification types, their performance will directly depend on the choice of modifications used to scan the test data. We ran many test trials with distinct sets of variable modifications to obtain averaged results for each algorithm. The Modi algorithm places no restriction on the number of variable modifications, but does require the database to be trimmed to no more than twenty proteins [6]. Thus, only one trial using the universal modification list and the twenty most abundant proteins was completed for Modi. PILOT_PTM demonstrates superior prediction accuracy for all scoring metrics and consistently outperforms the top hybrid and database algorithms.

[1] A. I. Nesvizhskii, O. Vitek, and R. Aebersold. Analysis and validation of proteomic data generated by tandem mass spectrometry. Nat. Methods, 4(11):787-797, 2007.

[2] E. S. Witze, W. M. Old, K. A. Resing, and N. G. Ahn. Mapping protein post-translational modifications with mass spectrometry. Nat. Methods, 4(11):798-806, 2007.

[3] D. Tsur, S. Tanner, E. Zandi, V. Bafna, and P. A. Pevzner. Identification of post-translational modifications by blind search of mass spectra. Nat. Biotechnol., 23(12):1562-1567, 2005.

[4] C. Baumgartner, T. Rejtar, M. Kullolli, L. M. Akella, and B. L. Karger. SeMoP: A New Computational Strategy for the Unrestricted Search for Modified Peptides Using LCMS/MS Data. J. Proteome Res., 7(9):4199-4208, 2008.

[5] S. Tanner, H. Shu, A. Frank, L. C.Wang, E. Zandi, M. Mumby, P. A. Pevzner, and V. Bafna. InsPecT: Identification of Posttranslationally Modified Peptides from Tandem Mass Spectrometry. Anal. Chem., 77(14):4626-4639, 2005.

[6] S. Kim, S. Na, J. W. Sim, H. Park, J. Jeong, H. Kim, Y. Seo, J. Seo, K. J. Lee, and E. Paek. Modi: a powerful and convenient web server for identifying multiple post-translational peptide modifications from tandem mass spectra. Nucleic Acids Res., 34:258-263, 2006.

[7] R. Matthiesen, M. Lundsgaard, K. G. Welinder, and G. Bauw. Interpreting peptide mass spectra by VEMS. Bioinformatics, 19(6):792-793, 2003.

[8] D. M. Creasy and J. S. Cottrell. Error tolerant searching of uninterpreted tandem mass spectrometry data. Proteomics, 2:1426-1431, 2002.

[9] R. Craig and R. C. Beavis. TANDEM: matching proteins with tandem mass spectra. Bioinformatics, 20(9):1466-1467, 2004.

[10] P. A. DiMaggio Jr. and C. A. Floudas. De Novo Peptide Identification via Tandem Mass Spectrometry and Mixed-Integer Optimization. AIChE Jour., 53(1):160-173, 2007.

[11] P. A. DiMaggio Jr. and C. A. Floudas. De novo peptide identification via tandem mass spectrometry and integer linear optimization. Anal. Chem., 79(4):1433-1446, 2007.

[12] P. A. DiMaggio Jr., C. A. Floudas, B. Lu, and J. R. Yates III. A Hybrid Method for Peptide Identification Using Integer Linear Optimization, Local Database Search, and Quadrupole Time-of-Flight or OrbiTrap Tandem Mass Spectrometry. J. Proteome Res., 7:1584-1593, 2008.

See more of this Session: Advances in Proteomics: New Technologies
See more of this Group/Topical: Topical 3: 2009 Annual Meeting of the American Electrophoresis Society (AES)

# Detecting Low Abundance Proteins in Skinned Cardiac Myofibrils

Rylie Siu, Department of Neurobiology, Physiology and Behavior, University of California, Davis, Davis, CA

Qian Xu, Department of Neurobiology, Physiology and Behavior, University of California, Davis, Davis, CA

Shannamar Dewey, Department of Neurobiology, Physiology and Behavior, University of California, Davis, Davis, CA

Susan Nguyen, Department of Neurobiology, Physiology and Behavior, University of California, Davis, Davis, CA

Aldrin V. Gomes,Department of Neurobiology, Physiology and Behavior, University of California, Davis, Davis, CA

Skinned cardiac myofibrils are the most commonly utilized muscle system to investigate muscle physiology. However, other than the well established components of the skinned myofibrils very little is known about the low abundance proteins present in skinned myofibrils. In 2005, our group found that a kinase was present in the skinned fiber which phosphorylated troponin I under certain conditions (Gomes et al. J. Mol. Cell. Cardiology 2005, 39:754-65). Western blotting has suggested the presence of the kinase cAMP dependent Protein kinase A (PKA) as well as the proteolytic complex, the proteasome, in the skinned myofibrils. Electrophoresis followed by trypsin digestion of gel pieces and LC-MS/MS is a powerful method for detecting proteins and associating partners (Gomes et al. Mol. Cell Proteomics 2009, 8: 302-315). One dimensional electrophoresis, trypsin digestion of the gel pieces and LC-MS/MS showed the presence of many proteins in skinned fibers but no kinases or proteasomes. The main problem is the large dynamic range of protein concentrations in the skinned fiber. Actin and myosin together make up more than 40% of the skinned fiber and dominate detection by mass spectrometry. To detect the medium to low abundant proteins the Proteominer protein enrichment kit was utilized. The Proteominer technology is based on the treatment of complex protein samples with a large, highly diverse library of hexapeptides bound to chromatographic supports. Each unique hexapeptide should bind to a unique protein sequence resulting in highly abundant proteins saturating their ligands, with excess protein being washed out and low abundance proteins being concentrated, ultimately reducing the dynamic range of the proteins in the sample. Few of the proteins present on cardiac myofibrils are released by homogenization in PBS so detergents (NP40, Triton X-100, SDS) are utilized to extract proteins. One problem with the use of hexapeptides is that they have not been optimized for samples containing detergents or urea. No method is currently available for using the hexapeptide beads on samples exposed to detergent. Isolation of the myofibrillar proteins in SDS containing solution followed by dialysis in PBS results in precipitation of many of the proteins. Several attempts to find a hexapeptide compatible medium that solubilize the myofibrillar proteins resulted in two suitable candidates: hexafluoroisopropanol (HFIP) and trifluoroethanol (TFE). HFIP and TFE are typically utilized to solubilize proteins. While both solvents were found to be useful: TFE and HFIP gave similar results, but TFE was found to be easier to use. After solubilization in SDS buffer, samples are precipitated with TCA/acetone and homogenized in Tris buffered TFE (40% v/v). This procedure resulted in most of the proteins being resolubilized in the Tris buffered TFE which was compatible with the hexapeptide beads. Porcine skinned fibers analyzed by the hexapeptide beads showed distinct changes in the banding pattern on one dimensional gels, suggesting that medium to low abundant proteins were enriched by this method. LC-MS/MS confirmed that medium to low abundant proteins were enriched by the Proteominer hexapeptide beads. It is also important to note that the high concentrations of proteins recommended for incubating the hexapeptide beads in the Proteominer manual (>50mg/ml) are not required for high protein retrieval from the beads.

See more of this Session: Advances in Proteomics: New Technologies
See more of this Group/Topical: Topical 3: 2009 Annual Meeting of the American Electrophoresis Society (AES)

# A Mismatch-Discriminating Open-Flow Carbon Nanotube Electrochemical DNA Sensor

**Sagnik Basuray**, Chemical and Biomolecular Engineering, University of Notre Dame, Notre Dame, IN

Satyajyoti Senapati, Chemical and Biomolecular Engineering, University of Notre Dame, Notre Dame, IN

Andrew R. Mahon, Center for Aquatic Conservation, Department of Biological Sciences, University of Notre Dame, Notre Dame, IN

Hsueh-Chia Chang, Chemical and Biomolecular Engineering, University of Notre Dame, Notre Dame, IN

The development of detection devices according to market demands needs to be rapid, sensitive, low cost and portable for easy accessibility to consumers. Conventional lab based assays although offer single molecule sensitivity, employ bulky and expensive fluorescent detection units that require qualified professional for their operations. Herein, we present an open-flow diagnostic chip that allows for the examination of a large volume (~100 microliter) sample with a small number of target molecules. The unit has a simple electrochemical impedance-based DNA biosensor with dielectrophoretic entrainment of multiwall carbon nanotubes (MWNT) across interdigital electrodes to produce maximum hydrodynamic shear force. The nanotube backbone is functionalized with a complementary probe to study DNA hybridization within a microfluidic chip. This is achieved by linking an amino functionalized single-stranded DNA probe onto carboxylated MWNTs by using water-soluble carbodiimide chemistry by crosslinking -NH2 and -COOH groups. Before and after hybridization with the complementary DNA sequences, a change of impedance is observed due to docking of DNA on the surface of CNT. It is observed that the flow within the chip creates a large enough shear force to differentiate three-base mismatches when compared with DNA from targeted species and a closely related congeneric species having mismatches at positions 12, 21 and 22 of the 26 base tag sequences (reading from 5' to 3', respectively). Typically in most electrochemical impedance devices with DNA tagging on electrodes shows a Randle circuit wherein the double layer capacitance dominates near the detection frequency range complicating the measurement of hybridization events. Previous studies implemented various novel methods to overcome this limitation. However, in our case, the impedance characteristics show a Warburg Impedance at a high frequency, without interference from the parasitic double layer thus allowing a quantitative and qualitative scrutiny of the hybridization-sensitive charge transfer reaction on the MWNT. The detection time for the device is only 20 minutes. Thus, we demonstrate for the first time, a rapid and sensitive detection of hybridization events at picomolar (pM) concentrations of target DNA with shear enhanced discrimination in a label-free CNT based microfluidic platform.

See more of this Session: Biomems and Microfluidics: Biomedical Diagnostics I
See more of this Group/Topical: Topical 3: 2009 Annual Meeting of the American Electrophoresis Society (AES)

# Microfluidic Capturing and Detection Dynamics of a Magnetic Nanoparticle Based Biosystem

**H. Susan Zhou**, Worcester Polytechnic Institute, Worcester, MA

Ahsan Munir, Worcester Polytechnic Institute, Worcester, MA

Jianlong Wang, Worcester Polytechnic Institute, Worcester, MA

Microfluidics combined with nanotechnology is playing a key role in transforming conventional biological work into Lab-on-chip devices; devices in which the entire analysis of a sample can be performed at the site of interest, including separation, and detection. These devices are now being realized for medical diagnostics in the hospital, forensic analysis of DNA at site of crime, and environmental analysis at a pollution site. Recently, Magnetic force methods are combined with microfluidic approach for on-chip separation and detection of biomolecules. These methods are based on manipulating the chemical or biological target molecules by attaching it to magnetic carriers and using an external magnetic field. However, most of the recent developments made in this area is based on functionalized magnetic beads or microparticles, there are relative few microfluidic systems developed that have employed magnetic nanoparticles. Compared with magnetic microparticles or microbeads, magnetic nanoparticles offer biocompatibility, stability, higher surface to volume ratio, minimum disturbance caused due to attached biomolecules, easily modifiable surfaces and moreover they are superparamagnetic.

While the selectivity and the specificity of attaching target molecule to magnetic carrier are related to the surface functionalization, the efficiency of the magnetic manipulation in microfluidics involves interplay of various other parameters such as inlet velocity of fluid containing magnetic nanoparticles, size of nanoparticles, magnetic field strength and its orientation, geometry of the device etc. In order to provide more quantitative comprehension of separation and detection of biomolecules using magnetic nanoparticles in microfluidic system and consequently help in designing, optimizing and developing more robust, efficient microfluidic lab-on-chip devices, finite element models are developed both for predicting the transport and capture of magnetic nanoparticles in a microfluidic system as well as for quantifying the effect of convection, diffusion, and magnetic field on the surface based detection.

The trajectories and trapping efficiencies were calculated and it was demonstrated that not only the size but also the point of release of nanoparticles within the microchannel affects the separation process. Influence of three important parameters, inlet velocities of fluid, diameter of magnetic nanoparticles and magnetic field strength on the trapping efficiency were investigated and optimized values of inlet velocity and magnetic field strength for the simulated system were predicted. It was further demonstrated that the angular position of magnet around the microchannel is also critical in dictating the resulting bioseparation performance. The efficiency of detection process greatly depends on kinetics and dynamics of surface-binding reaction between the target analyte and surface bound probes. The target molecules when tagged with magnetic nanoparticles were efficiently focused on the sensing site due to magnetic force which resulted in enhanced mass transport, faster binding, and consequently considerable reduction in detection time of target molecules when compared with physical micromixing.

The simulation performed using the developed models at the concept stage provided an excellent estimate of the potential to use magnetic nanoparticles for integrated surface-based bioassays. It can also help to investigate a wide range of design parameters and consequently predict an optimized and efficient design of these devices.

See more of this Session: Biomems and Microfluidics: Biomedical Diagnostics I
See more of this Group/Topical: Topical 3: 2009 Annual Meeting of the American Electrophoresis Society (AES)

# A Microfluidic Obstacle Course for Arraying and Detection of Microbeads, Lipobeads and Liposomes

Shahab Shojaei-Zadeh,Benjamin Levich Institute, New York, NY

David Bouttes, Ecole Polytechnique, Paris, France

Xiaohua Fang, Department of Earth and Environmental Engineering, Langmuir Center for Colloids and Interfaces, New York, NY

Bingquan Li, Department of Earth and Environmental Engineering, Langmuir Center for Colloids and Interfaces, New York, NY

Charles Maldarelli, Levich Institute and Chemical Engineering, The City College of New York, New York, NY

Ponisseril Somasundaran, Columbia University, New York, NY

Bead based microarrays are promising platforms for implementing the high throughput, multiplexed assaying of the binding interactions of biomolecules (for example the binding of antigens and antibodies, or the conjugation of membrane receptors with small molecule ligands). In these arrays, each bead contains a particular probe molecule on its surface, and a code to identify this probe. Particles with different probes are mixed and subsequently bound onto a surface in a regular array. The array is then incubated with a target, and the binding of the target to particular probe molecules is identified (usually by fluorescently labeling the targets and scanning to find luminescent beads). The bead code is then read to identify the probe and complete the multiplexed assay. The promise of this platform lies in the fact that by decreasing the size of the beads, ultra-mininaturized platforms capable of an increased number of binding assays can be constructed.

Bead arraying is essential to the formation of these bead based microarrays, and most research has focused on using the covalent binding of functional groups on the bead surface to functionalized sites on the platform surface. Arraying paradigms which sequester particles without reaction have the distinct advantage that they avoid chemical reaction conjugation. This presentation describes the design of a microfluidics cell for the placement, through flow, of micron-sized objects in a two dimensional array on a planar surface. The cell is fabricated out of poly(dimethylsiloxane) (PDMS) using soft lithography as a flow channel, with the dimensions of the channel approximately 50 microns in height and a few millimeters in length (i.e. along the flow direction) and width (perpendicular to the flow direction). Capture bodies project upwards from the floor of the channel and are approximately the same height as the channel. The capture bodies are arranged in rows perpendicular to the lengthwise (flow) direction of the channel; each capture body is separated from its neighbor in the row by 200 microns, and the rows are 200 microns apart. The capture bodies consist of three elements which are arranged to form a cup-shaped enclosure. The elements in each body are separated from one another by gaps of approximately 25 microns in width to allow fluid to flow through the cup along the channel direction. From row to row the bodies are staggered so that the gaps in each capture body arrange in approximately straight lines which are parallel to the lengthwise direction of the channel and insure approximately rectilinear streamlines through the obstacle course when the capture bodies are un-occupied. This obstacle course allows the enclosures of each capture body to retain objects that are suspended in an aqueous stream and allowed to flow from an inlet through the course. The elements of the capture body are curved to promote the deflection of particles not moving along the streamlines which intersect through the capture body. In this way the number of particles at each capture site is minimized.

The capture of spherical polymer beads, 40 microns in diameter, in this microfludic obstacle course is studied in order to construct a phase diagram of the statistics of bead occupancy as a function of the particle concentration, and the exposure time of the course to the flow. These studies are extended to the case of arraying liposomes and polymer beads with a lipid bilayer formed on the bead surface (lipobeads). Liposomes and lipobeads are more proper platforms for the display of membrane receptors for screening their interactions with target ligands. In this case, the capture bodies, floor and ceiling of the microfluidic cell are coated by a polyethylene glycol monolayer to prevent nonspecific adsorption of liposomes and lipobeads onto these surfaces. Retention of intact liposomes and lipobeads with intact bilayers are verified by confocal laser scanning confocal fluorescence microscopy of fluorescent labels sequestered in the bilayers.

See more of this Session: Biomems and Microfluidics: Biomedical Diagnostics I
See more of this Group/Topical: Topical 3: 2009 Annual Meeting of the American Electrophoresis Society (AES)

# Electrostatically Actuated Microvalves Fabricated with Soft Lithographic Techniques for Integrated Biological Microsystems

**Joshua D. Tice**, Department of Chemical & Biomolecular Engineering, University of Illinois at Urbana-Champaign, Urbana, IL

Amit V. Desai, Department of Chemical & Biomolecular Engineering, University of Illinois at Urbana-Champaign, Urbana, IL

Gregory A. Ten Eyck, Sandia National Laboratories, Albuquerque, NM

Christopher A. Apblett, Sandia National Laboratories, Albuquerque, NM

Paul J. A. Kenis, Department of Chemical & Biomolecular Engineering, University of Illinois at Urbana-Champaign, Urbana, IL

As efforts continue to create biological microsystems with higher degrees of integration, improving microvalve technology will be critical. "Quake"-style pneumatic microvalves have become prevalent in the field due to their straight forward soft-lithographic fabrication [1]. Thousands can be created in parallel and integrated into complex networks with high number density. However, these microvalves require several encumbering ancillaries, including a pressurized gas source, an array of solenoid valves, and computerized controls, which limit portability. Maharbiz et al. created an electrostatic microvalve with less cumbersome ancillaries that retains similar performance characteristics [2, 3]. The tradeoff is a complex fabrication process that includes high-temperature annealing, photoresist patterning, metal evaporation, and etching.

We report the design, fabrication, and testing of an electrostatically actuated microvalve made exclusively with soft-lithographic techniques. Our novel fabrication significantly reduces the complexity of fabrication and integration. The valve comprises an elastomeric membrane embedded with conducting nanoparticles and suspended above a lower electrode with the aid of a support structure. We have developed an analytical model that identifies several parameters affecting actuation potential, including distance between the electrodes, diameter of the microvalves, and elastic modulus of the elastomer. Additionally, the shape of the membrane's support structure influences actuation potential. We will present work to optimize the valve design by altering the geometry of the support structure, a comprehensive study of valve isolation pressures with different fluids, and integration of the valve into a biomedical device.

REFERENCES

1. M. A. Unger, H. P. Chou, T. Thorsen, A. Scherer and S. R. Quake, Science 2000, 288, 113-116.

2. T. Bansal, M. P. Chang and M. M. Maharbiz, Lab Chip 2007, 7, 164-166.

3. M. P. Chang and M. M. Maharbiz, Lab Chip 2009, 9, 1274-1281.

See more of this Session: Biomems and Microfluidics: Biomedical Diagnostics I
See more of this Group/Topical: Topical 3: 2009 Annual Meeting of the American Electrophoresis Society (AES)

# Femtomolar Detection for Microfluidic Immunoassays Using Controlled Evaporation

**Nicholas S. Lynn**, Department of Chemical and Biological Engineering, Colorado State University, Fort Collins, CO

Charles S. Henry, Department of Chemistry, Colorado State University, Fort Collins, CO

David S. Dandy, Department of Chemical and Biological Engineering, Colorado State University, Fort Collins, CO

Diagnostic protocols based on protein-antigen affinity interactions have become increasingly popular in both clinical and research settings. Heterogeneous affinity assays, in which antigens (or antibodies) immobilized to a surface capture analytes of interest from an aqueous solution, offer many advantages over their traditional homogeneous well-based counterparts, including reductions in footprint size, analysis times, and sample volumes. However, because these methods rely on the reversible protein-antigen affinity interaction, the equilibrium analyte coverage is a function of the association constant for a given system. As a result, affinity interactions possessing poor association kinetics are restricted to limits of detection (LOD) above 1 pM when using standard techniques of detection (e.g. SPR or fluorescence assays).

In this work we present a method that utilizes the effects of evaporation to passively concentrate an analyte solution to lower the LOD for heterogeneous assays. A microfluidic system composed of two reservoirs connected by a single microchannel is reversibly bonded to a substrate such that an outlet reservoir lies over discrete regions of immobilized antibodies. Upon introduction of analyte solution into the inlet reservoir, fluid spontaneously fills microchannel and the corner regions of an outlet reservoir such that the liquid volume present in the outlet reservoir quickly establishes a steady state value on the sub-nL scale and liquid is passively pumped from the inlet to outlet reservoir. With optimized design of the microchannel and outlet reservoir geometries, the ratio of the volumetric flow rate through the channel to the steady state liquid volume can be increased to the point where the liquid solution in the outlet reservoir is concentrated by over three orders of magnitude in less than 30 minutes. Because the liquid volume in the outlet reservoir is steady over time, unwanted deposition of material onto the substrate is avoided.

We show that this passive system can be used to lower the LOD for heterogeneous immunoassays by multiple orders of magnitude while requiring no external pumping mechanism, and eliminates the need for expensive detection methods or complex fabrication strategies.

See more of this Session: Biomems and Microfluidics: Biomedical Diagnostics I

See more of this Group/Topical: Topical 3: 2009 Annual Meeting of the American Electrophoresis Society (AES)

# Microfluidic Chips for Multiplexed Viral DNA Detection

**Benjamin R. Schudel**, Department of Chemical & Biomolecular Engineering, University of Illinois at Urbana-Champaign, Urbana, IL

Melikhan Tanyeri, Department of Chemical & Biomolecular Engineering, University of Illinois at Urbana-Champaign, Urbana, IL

Charles M. Schroeder, Department of Chemical & Biomolecular Engineering, University of Illinois at Urbana-Champaign, Urbana, IL

Paul J. A. Kenis, Department of Chemical & Biomolecular Engineering, University of Illinois at Urbana-Champaign, Urbana, IL

Molecular beacons are a highly sensitive means of detection of DNA sequences with specificity down to the single base pair. To date, this detection concept has not been used in high throughput screening, due to the expense associated with these reagents. Microfluidic devices, however, use mere nano- or picoliter volumes of sample, thus overcoming the cost issue and opening up the use of these highly sensitive molecular beacons for high throughput applications.

Here we present the use of a microfluidic chip to create a combinatorial array of multiplexed molecular beacons against a variance of oligomer targets. This chip utilizes Actuate-to-Open valves which allow for easy filling and isolation of the compartments [1]. Furthermore, the valves do not need to be actuated continuously after filling, so the chip can be disconnected from the ancillary pressure source and transported for further use or analysis.

Biotinylated molecular beacons with different probe regions are specifically bound to a streptavidin-patterned surface in different compartments. Total internal reflection fluorescence (TIRF) will be used to detect whether or not certain viral targets are present across an array of beacons. The key advantage of TIRF in combination with molecular beacons is that the presence of very low concentrations of oligomer target can be detected. This has potential application in point-of-care testing in clinical settings.

See more of this Session: Biomems and Microfluidics: Biomedical Diagnostics I
See more of this Group/Topical: Topical 3: 2009 Annual Meeting of the American Electrophoresis Society (AES)

# Single-Molecule Target Sequence Detection Using Extensional Flow

Rebecca Dylla-Spears, Chemical Engineering, University of California at Berkeley, Berkeley, CA

Lydia L. Sohn, Mechanical Engineering, University of California at Berkeley, Berkeley, CA

Susan J. Muller, Chemical Engineering, University of California at Berkeley, Berkeley, CA

We use planar extensional flow to trap and extend single molecules of double-stranded genomic DNA for detection of target sequences along the DNA backbone. Fluorescently tagged enzymatic probes are bound sequence-specifically to fluorescently labeled double-stranded λ-DNA. The complexes are introduced into a microfluidic cross slot, where the tagged DNA molecules are trapped and elongated at the stagnation point of the planar extensional flow. The degree of elongation can be controlled using the flow strength in the device. Beads bound along the stretched DNA may be directly observed, and their locations along the backbone determined, using fluorescence microscopy. Previously, we have demonstrated the feasibility of this method for detection of specific sequences, for single-molecule kinetic and binding studies, as well as for single-molecule sorting (R. Dylla-Spears, L.L. Sohn, and S.J. Muller, AIChE Annual Meeting, Salt Lake City, November 2007). We now compare the measured bead positions to the expected EcoRI target sequences on λ-DNA to determine the method's precision and accuracy at molecular extensions of 70% and 90% of the contour length.

See more of this Session: Biomems and Microfluidics: Biomedical Diagnostics I
See more of this Group/Topical: Topical 3: 2009 Annual Meeting of the American Electrophoresis Society (AES)

# Integrated AC Electrokinetic Microdevices

**Zachary R. Gagnon**, Chemical and Biomolecular Engineering, University of Notre Dame, Notre Dame, IN

Hsueh-Chia Chang, Chemical and Biomolecular Engineering, University of Notre Dame, Notre Dame, IN

Manipulating the most basic lab-on-a-chip (LoC) ingredients - particles, cells and fluid - in closed micro-channel geometries is a requirement of any LoC-type device. Specific tasks - fluid pumping, sample mixing, bioparticle concentration, separation, and analyte detection - require micro-sized pumps, mixers, concentrators, separators and sensors. Conventional solutions to this problem involve scaling down peristaltic, diaphragm, and other popular macro-scale mechanical pumps or micromanipulators towards micro-scale dimensions. However, most such devices contain moving parts which can lead to clogging and cell lyses, and render them unreliable and unsuitable for biological applications.

Due to such drawbacks, the use of electrokinetic (EK) devices with no moving parts to transport, mix fluid and manipulate cells and particles has attracted considerable attention. The term Electrokinetic is used to describe the motion of a liquid or particle under the influence of an external microelectrode-driven electric field. EK phenomenon can be quite powerful at the micro-scale; the well developed integrated circuit (IC) fabrication techniques from the microprocessor industry have made the fabrication and assembly of microelectrodes into microfluidic geometries a simple reality. This allows the experimentalist to deliver highly spatially controlled alternating current (AC) or direct current (DC) electric fields and integrate electrokinetics into microfluidic devices for specific diagnostic tasks.

The design, control, and implementation strategies, however, of EK devices for use in LoC applications, specifically in biological microfluidic applications, can be non-trivial and exceedingly challenging. Here we present the discovery, development and optimization of such strategies and their integration into a fully functional portable LoC device for sample control and manipulation. Specifically we discuss the integration of micro-electrode induced AC electrokinetic phenomena such as dielectrophoresis and AC electro-osmosis within microfluidic systems for use in biological and nanoscale lab-on-a-chip applications ranging from genetic detection to yeast cell and malaria infected blood cell characterization and sorting. It will be shown that such integrated electrokinetic effects satisfy the major requirements of next generation microfluidic diagnostics - namely fluid metering, sample concentration, and detection and quantification.

See more of this Session: Biomems and Microfluidics: Biomedical Diagnostics II
See more of this Group/Topical: Topical 3: 2009 Annual Meeting of the American Electrophoresis Society (AES)

# DIELECTROPHORETIC MANIPULATION OF LIVER CANCER CELLS

*Lucía D. Garza-García, Marcos Garza-Madrid, Sergio Serna-Saldívar, Blanca H. Lapizco-Encinas**

*Tecnologico de Monterrey, Monterrey, N.L., Mexico*

## Abstract

Dielectrophoresis is an efficient microscale technique employed for the manipulation and separation bioparticles, with great potential for clinical applications. The present study demonstrates the manipulation and concentration of live Hep-G2 cells by employing insulator-based dielectrophoresis (iDEP) and DC electric fields. The fluorescence stained Hep-G2 cells were introduced in a plastic microchannel, which had an array of cylindrical insulating posts. The results showed that the magnitude of the applied electric field and the suspending medium properties have a great impact in the dielectrophoretic response of the Hep-G2 cells.

## Introduction

Lab on a chip devices are becoming increasingly important in many biomedical applications. Microfluidics offers important advantages that are critical in the biomedical field: fast response time, portability, low sample and reagent consumption. One of the separation techniques with great potential to be used in microdevices is dielectrophoresis (DEP). DEP has been successfully employed for the manipulation of biomolecules and cells.[1,2]

DEP is the movement of particles in suspension, due to polarization effects, when particles are exposed to a non-uniform electric field.[3] When a particle is exposed to an electric field, the particle becomes polarized, a dipole is induced in the particle, and the magnitude and direction of this dipole depends on the characteristics of the electric field (magnitude and frequency) as well as on the dielectric properties of the particle and suspending medium.[4] If the electric field is spatially uniform, the Coulombic attractions between the dipolar charges and the electric field are balanced and there is no net movement, unless the particle has a charge and the field is direct current or has a low frequency. However, when the electric field is spatially no-uniform, there is an unbalance between the Coulombic forces, leading to net movement of the particle in the direction where the field magnitude is higher. Since the spatial variation of the field governs particle motion, the particle will always move along the direction in which the electric field has a greater gradient. This particle motion is independent on electric field polarity, therefore, DEP can take place in alternating current (AC) or direct current (DC) electric fields.[4] DEP can be classified by the direction of particle motion. If the particle moves toward the regions of increasing field gradient, then it is termed as positive dielectrophoresis, which occurs when the particle has a higher polarizability than the suspending medium. On the contrary, negative dielectrophoresis occurs when the particle is less polarizable than the medium and the particle motion is toward the regions of decreasing field gradient.[5]

There are two main approaches for the generation of non-uniform electric fields: employing arrays of microelectrodes or employing arrays of insulting structures. Most of the studies reported on DEP have been performed employing arrays of microelectrodes.[6] However; electrode-based DEP has important drawbacks, since electrodes lose functionality due to fouling, which is a common effect when employing bioparticles. Additionally, fabrication of electrode-based systems is expensive and requires complex steps due to metal deposition processes.[7]

In insulator-based-DEP (iDEP) the electric field is applied employing only two electrodes that straddle an array of insulating structures. When an electric field is applied across an insulating structures array, the presence of the structures creates regions of higher and lower field strength, *i.e.* dielectrophoretic traps.[7] Insulator-based DEP allows for more robust systems, since insulators can function despite of fouling, and can be fabricated from a variety of material, such as plastics, making it possible for inexpensive and disposable microdevices, ideal for biomedical applications.[8]

There have been several studies involving DEP applied to mammalian cells, demonstrating the effectiveness of this technique for cell concentration and sorting. For instance, in 1994 Gascoyne *et. al.*[9] reported the separation of cancerous and normal cells, employing a dielectrophoretic collection spectra with an interdigitated gold electrode array. In a later study, the same group[10] reported the application of dielectrophoretic field-flow fractionation to separate circulating tumor cells from clinical blood specimens. Ho and collaborators[11] demonstrated a biomimetic heterogeneous patterning of hepatic and endothelial cells through the application of positive DEP in concentric star-like array of electrodes. Sankaran *et. al.*[12] captured NIH 373 fibroblast cells with DEP, using transparent indium tin oxide microelectrodes. Jen and Chen[13] demonstrated the trapping of live and dead HeLa cells by applying iDEP within open-top microstructures.

The present work reports the manipulation and concentration of liver cancer cells (Hep-G2) cells, employing iDEP and DC electric fields. The experiments were carried out in a plastic microchannel, which had an array of cylindrical insulating posts. The results obtained demonstrated that liver cancer cells can be immobilized with DEP, that the dielectrophoretic response of Hep-G2 cells can be controlled by manipulating the magnitude of the applied electric field and suspending medium properties (pH and conductivity).

## Theory of dielectrophoresis

As mentioned, DEP can only be achieved in the presence of a nonuniform electric field. The DEP force exerted on a spherical particle is represented with the following equation:

$$F_{DEF} = 2\pi\varepsilon_0\varepsilon_m r_p^3 f_{CM} \nabla E^2 \qquad (1)$$

where $\varepsilon_0$ is the permittivity of the free space, $\varepsilon m$ is the relative permittivity of the suspending medium, *rp* is the particle radio, *E* is the magnitude of the applied electric field, and *fCM* is the Clausius-Mossotti (CM) factor, also called polarization factor. The CM can be estimated in terms of the real conductivity of the particle and the medium when DC electric fields are applied:[14]

$$f_{\mathrm{CM}} = \left[\frac{\tilde{\sigma}_p - \tilde{\sigma}_m}{\tilde{\sigma}_p + 2\tilde{\sigma}_m}\right] \qquad (2)$$

where $\tilde{\sigma}_p$ and $\tilde{\sigma}_m$ are the complex conductivities of the particle and medium respectively. Therefore, a positive value of *CM* means of positive DEP, and vice versa. Considering these two equations, one can see that the DEP force exerted on a particle depends on the characteristics of the applied field, particle and suspending medium properties. These operating conditions can be optimized in order to achieve effective particle concentration and separation.

When DC fields are applied, electroosmotic flow and electrophoresis are present. These two phenomena have a linear dependency on electric field. Dielectrophoresis is of second order with respect to the applied electric field. When the electric field applied is low, low electric field gradients will be produced as a consequence, leading to negligible dielectrophoretic force. In order to achieve dielectrophoretic particle trapping or immobilization, DEP has to be the governing force present in the system, *i.e.*, DEP has to overcome electroosmotic flow and electrophoresis. When handling mammalian cells, which are large, electrophoresis can be neglected.

## Materials and Methods

### Microdevice

The device employed was made from zeonor polymer[8] with eight microchannels; each channel was 1-mm wide, 75-μm deep, and 10.2-mm long. Each microchannel contained an array of cylindrical insulating posts that had a diameter of 150 μm and were arranged 250 μm center-to-center. A dove-tail geometry was added to the first row of the post in each side, in order to prevent cells from colliding against the posts and obstructing the system. The post array had 12 columns of 5 posts each, and was straddled by an inlet and outlet reservoirs.[8] Electrodes were placed at the reservoirs in order to apply the electric field across the microchannel length (Figure 1).

### Cells

Hep-G2 cells were obtained from the ATCC cell bank and cultured on Petri dishes using Dulbecco Modified Eagle Media (DMEM, Sigma-Aldrich, USA) supplemented with 5% FCS (Sigma-Aldrich, USA). Cells were harvested by incubation in 2.5% Tripsin (Sigma-Aldrich, USA) for 10 min. $1x10^7$ cells were transferred to a clean centrifuge tube and centrifuged at 1,000 RPM for 5 minutes. Finally the media was removed by careful aspiration and the cells were resuspended in 10 mL of suspending medium for a final concentration of $1x10^6$ cells/mL. For the visualization in the microscope, Hep-G2 cells were stained employing SYTO 9 (Invitrogen, Carlsbad, CA, USA), a DNA-intercalating green fluorescent dye (excitation/emission 483/503 nm). Per manufacturer instructions, 3 μL of SYTO 9 were added to a milliliter of the suspended cells. The cells were then incubated with the dye for 15 minutes in the dark at room temperature.

### Equipment

DC electric fields were applied employing a 3000 V high voltage sequencer, model HVS448 (LabSmith, Livermore, CA, USA). In order to record the dielectrophoretic response of the Hep-G2 cells, an inverted epifluorescence optical video microscope for microfluidics and microbiology, model SVM340 (LabSmith, Livermore, CA, USA), was used. All of the experiments were recorded employing a 10X microscope objective.

### Suspending medium

In order to protect Hep-G2 cells from osmotic shock and to maintain them alive for a longer period of time, they were suspended in a 5% dextrose solution (Laboratorios PISA, Mexico) with a pH of 7.4 and a conductivity of 100 μS/cm. After staining, 40 μL of cell solution were introduced at the inlet reservoir of the microchannel and the electrodes were positioned at the inlet and outlet reservoirs. The dielectrophoresis force was observed when a high voltage was applied across the microchannel.

## Results and discussion

A set of experiments was carried out to evaluate the dielectrophoretic response of Hep-G2 cells, the results obtained are shown in Figure 2. From Figure 2a it can be observed that some of the Hep-G2 cells were immobilized by dielectrophoresis when a field of 150 V/cm was applied. The number of trapped Hep-G2 cells increased with the applied field magnitude, as shown in Figures 2b and 2c where fields of 300 V/cm and 400 V/cm were applied, respectively. These results demonstrate that dielectrophoretic immobilization of Hep-G2 cells increases with the applied field.

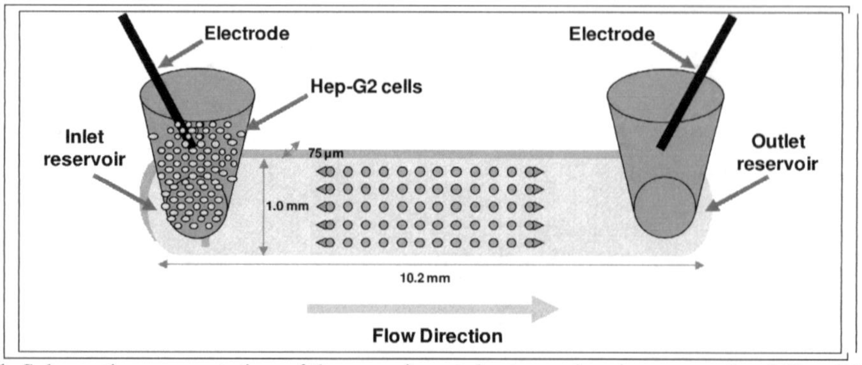


**Figure 1**. Schematic representation of the experimental set-up, showing a sample of Hep-G2 cells at the inlet reservoir.

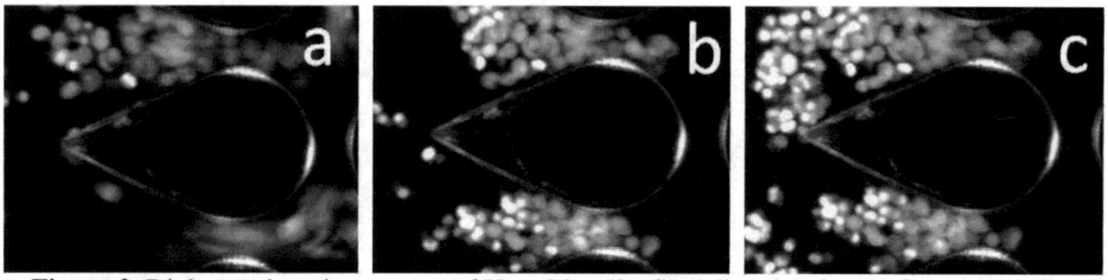


**Figure 2**. Dielectrophoretic response of Hep-G2 cells, flow direction from left to right. Electric fields applied are: (a) 150 V/cm, (b) 300 V/cm, (c) 400 V/cm. Cells were suspended in a 5% dextrose solution with a pH of 7.4 and $\sigma_m$=100 µS/cm.

The 5% dextrose solution with pH = 7.4 and $\sigma_m$=100 µS/cm resulted in an effective suspending medium; since the Hep-G2 cells were maintained alive for about 45 minutes under the presence of an electric field. After this time we observed the presence of detritus while most of the Hep-G2 cells shrank and showed reduced fluorescence, changes that correspond to membrane instability and apoptosis.

## Conclusions

This investigation demonstrates experimentally the manipulation and concentration of Hep-G2 cells by using iDEP and DC electric fields. Hep-G2 cells can be trapped with positive dielectrophoresis, where a higher electric field will cause a larger number of cells to be trapped. In order to reduce the stress of the Hep-G2 cells under the electric field applied, a dextrose solution at 5% with a low conductivity was used as the suspending medium. In the future, iDEP could be used in microdevices to identify and isolate cancer cells from a real patient sample, differentiating them from normal cells and opening the possibility for iDEP to be used as a tool for cancer diagnostics.

## Acknowledgements

The authors wish to thank the financial support given by grant CONACYT-CB-2006-53603, and *Cátedras de Investigación* of *Tecnológico de Monterrey* (CAT 142, CAT 122 and CAT 151).

## References


1 Voldman J., (2006), "Electrical forces for microscale cell manipulation," *Ann. Rev. Biomed. Eng.* 8, pp.425-454.

2 Lapizco-Encinas B.H., Rito-Palomares M., (2007), "Dielectrophoresis for the manipulation of nanobioparticles," *Electrophoresis* 28(24), pp.4521-4538.

3 Pohl H.A., (1951), "The Motion and Precipitation of Suspensoids in Divergent Electric Fields," *J. Appl. Phys.* 22(7), pp.869-871.

4 Hughes M.P., (2002), "Strategies for dielectrophoretic separation in laboratory-on-a-chip systems," *Electrophoresis* 23, pp.2569-25582.

5 Hughes M.P., (2002), "Nanoelectromechanics in Engineering and Biology," CRC Press, Boca Raton, FL.

6 Gascoyne P.R.C., Vykoukal J., (2002), "Particle separation by dielectrophoresis," *Electrophoresis* 23(13), pp.1973-1983.

7 Cummings E.B., Singh A.K., (2003), "Dielectrophoresis in Microchips Containing Arrays of Insulating Posts: Theoretical and Experimental Results," *Anal. Chem.* 75(18), pp.4724-4731.

8 Simmons B.A., McGraw G.J., Davalos R.V., Fiechtner G.J., Fintschenko Y., Cummings E.B., (2006), "The Development of Polymeric Devices as Dielectrophoretic Separators and Concentrators," *MRS Bulletin* 31, pp.120-124.

9 Gascoyne P.R.C., Noshari J., Becker F.F., R. P., (1994), "Use of Dielectrophoretic Collection Spectra for Characterizing Differences between Normal and Cancerous Cells.," *IEEE Trans. Ind. Appl.* 30, pp.829-834.

10 Gascoyne P.R.C., Noshari J., Anderson T.J., Becker F.F., (2009), "Isolation of Rare Cells from Cell Mixtures by Dielectrophoresis.," *Electrophoresis* 30(8), pp.1388-1398.

11 Ho C.-T., Lin R.-Z., Chang W.-Y., Chang H.-Y., Liu C.-H., (2006), "Rapid heterogeneous liver-cell on-chip patterning via the enhanced field-induced dielectrophoresis trap," *Lab Chip* 6, pp.724-734.

12 Sankaran B., Racic M., Tona A., Rao M.V., Gaitan M., Forry S.P., (2008), "Dielectrophoretic capture of mammalian cells using transparent indium tin oxide electrodes in microfluidic systems," *Electrophoresis* 29, pp.5047–5054.

13 Jen C.P., Chen T.W., (2008), "Selective trapping of live and dead mammalian cells using insulator-based dielectrophoresis within open-top microstructures," *Biomed. Microdev.* 11(3), pp.597-607.

14 Lapizco-Encinas B.H., Simmons B.A., Cummings E.B., Fintschenko Y., (2004), "Dielectrophoretic Concentration and Separation of Live and Dead Bacteria in an Array of Insulators," *Anal. Chem.* 76(6), pp.1571-1579.

# DC Dielectrophoresis: Separation of Fluorescent Polystyrene Particles

**Soumya K. Srivastava**, Mississippi State University, Mississippi State, MS

Adrienne Minerick, Mississippi State University, Mississippi State, MS

Blanca H. Lapizco-Encinas, Biotechnology and Food Engineering and Biotechnology Center, Tecnologico de Monterrey, Monterrey, Mexico

This research describes a lab-on-a-chip device for the separation of microparticles by size via Direct Current Dielectrophoresis (DC-DEP). The separation was achieved by using a rectangular insulating obstacle in a 200 micron channel to create non-uniformities in field density. Particles travel into one of four exit channels based on their deflection from the insulating obstacle. The device and insulating obstacle were fabricated with PDMS device by soft-photolithography process. The PDMS device was sealed to glass slide by various techniques. A baseline EOF was established by current monitoring method for the devices sealed by these techniques: 1) IPA treated glass slide with heat at 65 deg C 2) UV/Ozone treated PDMS device with DI water washed glass slide 3) electrostatic seal with no treatment. The DC-DEP particle flow behaviors were investigated with three different sizes of fluorescent polystyrene particles- 2.28 μm, 5.49 μm and 15.68 μm. 0.14 M phosphate buffer saline was used as a dilution buffer for all experiments. Field dependency studies were carried out at 3.42, 6.85 and 10.27 V/cm. Reproducibility was tested by repeating experiments at least 5 times or until the standard deviation was within 5% of the mean. An Axiovision CCD camera was used to record fluorescent images at 5 mS interval for 2 minutes. The data analysis was carried out by taking fluorescent intensity profiles at locations prior to and just past the insulating obstacle. These intensity profiles identify particle flow events in the channel and each position is recorded with time. The total intensity for all the events in the 2-minute experiment at a particular Y position were plotted to determine the trajectory of the particle. This work is an essential first step in using DC-DEP for particle sorting in a continuous flow lab on a chip device.

See more of this Session: Biomems and Microfluidics: Biomedical Diagnostics II
See more of this Group/Topical: Topical 3: 2009 Annual Meeting of the American Electrophoresis Society (AES)

# THE INTEGRATION OF DIELECTROPHORESIS ON A CD-LIKE MICROFLUIDICS PLATFORM

*Rodrigo Martinez-Duarte, Robert A. Gorkin III, Kameel Abi-Samra and Marc J. Madou*
*University of California, Irvine, Irvine, CA, USA*

## Introduction

Dielectrophoresis (DEP) enables the selective manipulation of a targeted particle, or population of particles, using the interaction of a non-uniform electric field with the induced effective dipole moment of the particle(s) (1). DEP is a promising technique that has been demonstrated in the selective manipulation of biological particles such as DNA (2-10), proteins (11, 12), microorganisms (13-17), bacteria (18-22), viruses (23-26) and cells (27-32). DEP is advantageous over other separation techniques because particle separation is solely based on the difference between the dielectric properties of the targeted particle and those of the rest of the sample. The dielectric properties of a particle are determined by physical properties, for example membrane morphology and internal compartmentalization in the case of cells, and give a particle a characteristic dielectric phenotype that is different from other particles and the surrounding media. The use of DEP for particle separation eliminates the need for specific labels, *e.g.*, magnetic beads or fluorophores as needed by other separation techniques such as flow cytometry and magnetophoresis. The elimination of these often expensive labels reduces the complexity and cost of an assay which can expand the availability of diagnostic tests, such as for HIV, and make cell therapies for cancer or degenerative diseases available to a broader number of patients. For further reading an excellent review on DEP fundamentals and the use of DEP in diagnostics is by Gascoyne and Vykoukal (33).

However, even when assay cost and complexity are reduced with the elimination of labels, the fact that a millimeter-sized DEP chip must be interfaced to large and often expensive precision apparatus for fluid dispensing has prevented DEP from becoming a practical separation platform. Furthermore, traditional DEP platforms rely on a combination of syringes, tubing, valves, and fluidic ports for fluid manipulation that significantly increases the complexity of the platform. In this work we replaced the traditional syringe piston pumping by centrifugal pumping on an electrically-interfaced Compact Disk (CD)-like microfluidic platform. Dielectrophoresis is carried on using modular DEP chips that are inserted in the CD platform. The inclusion of the entire fluid network in a single DEP chip eliminates the use of syringes, external valves, tubing and any other supporting fluidic elements. This work differs from DEP platforms where electroosmosis is used as pumping force in that centrifugal pumping does not depend on the electric nor physicochemical properties of the sample.

Microfluidic CD technology has been developed as a unique alternative for handling biological analysis (34-36). CD microfluidics rely on forces and pressure gradients induced on a spinning CD platform to move sample liquids. In its most basic form, fluids are propelled towards the outer radius of the disk by a centrifugally induced pressure gradient. The advantages of centrifugal pumping compared to other standard propulsion techniques such as syringe pumps, ultrasonic, or electrokinetic pumping have been well documented in Madou et al. (34). Key benefits of centrifugal pumping described in the latter reference are insensitivity to most pysicochemical properties of the sample being pumped (such as pH, ionic strength or chemical composition) and amenability to miniaturization and assay multiplexing.

The ultimate goal is to implement an automated sample-to-answer system where human intervention is only needed to load a sample from a patient, *e.g*, blood, saliva or urine, choose the assay to be performed and retrieve results, for example the number of CD4+ T-cells used for the monitoring of HIV patients. In specific cases, such as in cell theraphy, further human or robotic intervention will be required to retrieve the isolated cell fractions for further processing or implantation. In the current paper we clarify the fabrication process and describe the initial proof-of-concept of the platform for the selective filtering of yeast cells from a latex-yeast mixture. This work represents an important step towards the final goal of integrating all types of electrokinetic phenomena (AC and DC) to other functions we already demonstrated on CD platforms including valving, decanting, calibration, mixing, metering, sample splitting, and fluid separation.

The combination of centrifugal and electric forces in a single platform has been reported before by Wang and colleagues who demonstrated a dual-pumping CD-like electrophoresis-based separation system(37); by Fuhr and co-workers (38, 39) who disclosed a swinging-rotor centrifugal system to conduct DEP with metal-electrodes and by Boettcher and colleagues who presented a system somewhat similar to the one presented here (40).

## Platform

The experimental setup consists of five parts: 1) a programmable motorized spin stand, 2) a function generator, 3) a specialized spin chuck, 4) a compact disk (CD) platform and 5) DEP chips. Assembly of the system is illustrated in Fig. 1. The CD platform is mechanically connected to the motorized spin stand through a custom-made spin chuck that features a slip ring arrangement which enables the electrical connection of the spinning platform to a function generator. An electrical network is embedded on the CD platform to interface the signal coming from the function generator to the DEP chips. We use 3D carbon-electrode DEP, or carbonDEP, chips to demonstrate the platform. The carbonDEP chips include carbon 3D electrodes (70 μm-high) fabricated using the C-MEMS technique (41), and a microfluidic network fabricated with machined polycarbonate and double-sided pressure-sensitive adhesive. The channel containing the 70 μm-high carbon electrodes has a rectangular cross-section (600 μm width and 100 μm height). The fabrication process of carbonDEP chip and the experimental device are shown in Fig. 2 but detailed elsewhere (42-44). Briefly, a two-step photolithography process is used to fabricate SU-8 structures which are then pyrolyzed (heated at high temperatures in an inert atmosphere) to obtain carbon electrodes and their connecting electrical leads. A thin layer of SU-8 is photo-patterned around the electrodes and on top of the connecting leads to protect them from experimental reagents that can cause them to lift off from the substrate. Modular carbonDEP chips are placed in the CD platform using designated slots. This modular arrangement allows for an user-friendly and rapid experimentation setup. Furthermore, it enables the use of other DEP chips featuring different electrode materials and geometries.

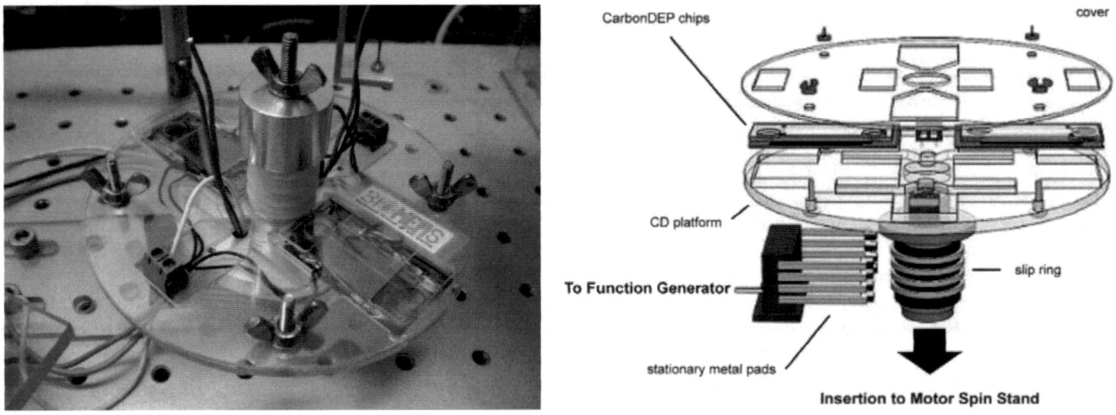


**Fig. 1** A CD-like centrifugal microfluidics platform for dielectrophoresis

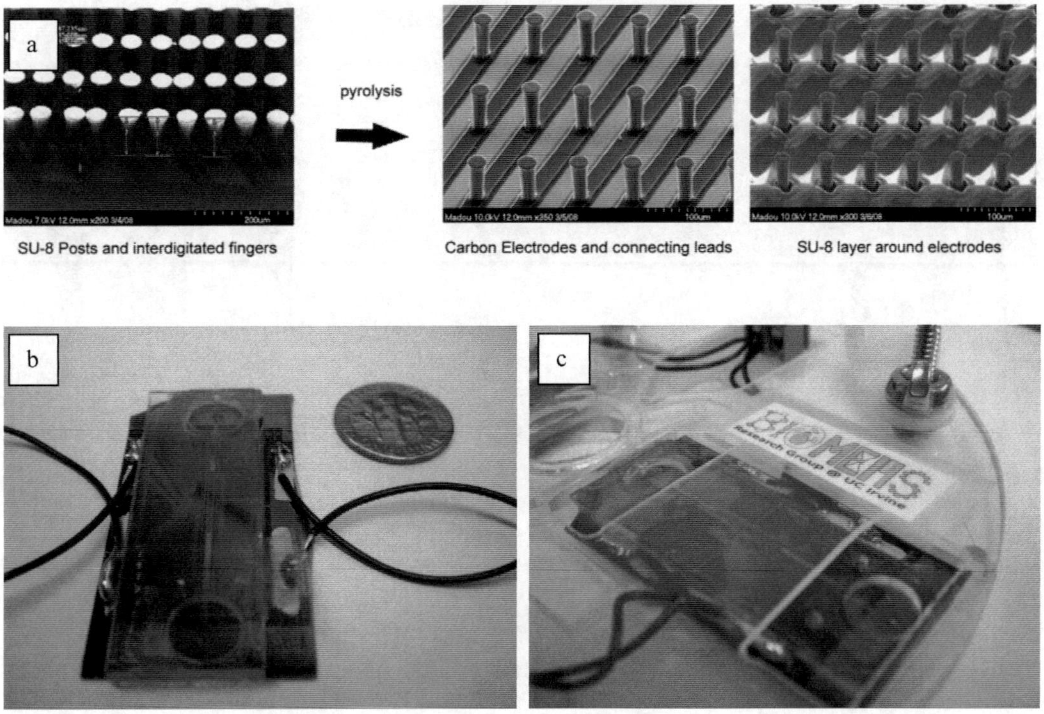


**Fig. 2** a) Fabrication process of carbon electrodes, b) experimental device compared in size to a U.S. dime and c) experimental device mounted in the CD platform

## Experimental

The experimental sample is a mixture of 8 μm latex particles and viable and non-viable yeast cells suspended in a DI water-based 0.1% Bovine Serum Albumin (BSA) solution. The sample features a pH of 4.4, a conductivity ($\sigma$) of 31.2 μS/cm and a particle concentration of 2.75 X $10^5$ particles per ml with 60% latex and 40% yeast cells. The working frequency and amplitude of the polarizing AC signal is 200 kHz and 20 $V_{pp}$ respectively. Our DEP models of the experimental sample (results not shown) suggest that at 200 kHz both non-viable and viable yeast cells are attracted to the electrode surface. At that same frequency, the latex particles are repelled from the electrodes and eluted by the fluid flow. In other words, at 200 kHz polarizing frequency the carbon electrode array acts as a dielectrophoresis-based active filter to filter out yeast cells from latex particles.

The experimental protocol is illustrated in Fig. 3. A 50 μl volume of the experimental sample is first introduced into the inner chamber (the chamber closest to the CD center of rotation). The sample is prevented from flowing into the channel containing the electrodes during loading because of the inner chamber-channel interface which acts as an hydrophobic valve (35). The electrodes are polarized immediately after sample loading using the AC signal detailed above. The channel containing the electrodes is then primed with the sample solution by rapid acceleration of the CD up to 800 rpm which causes the hydrophobic valve to burst (35). Once the channel is primed, the platform is set to spin at a constant speed to induce a desired flow rate. After the whole sample volume is consumed over the electrodes, the platform is stopped and the sample is retrieved from the outer chamber (the chamber closest to the CD edge). Electrodes are de-polarized after sample retrieval.

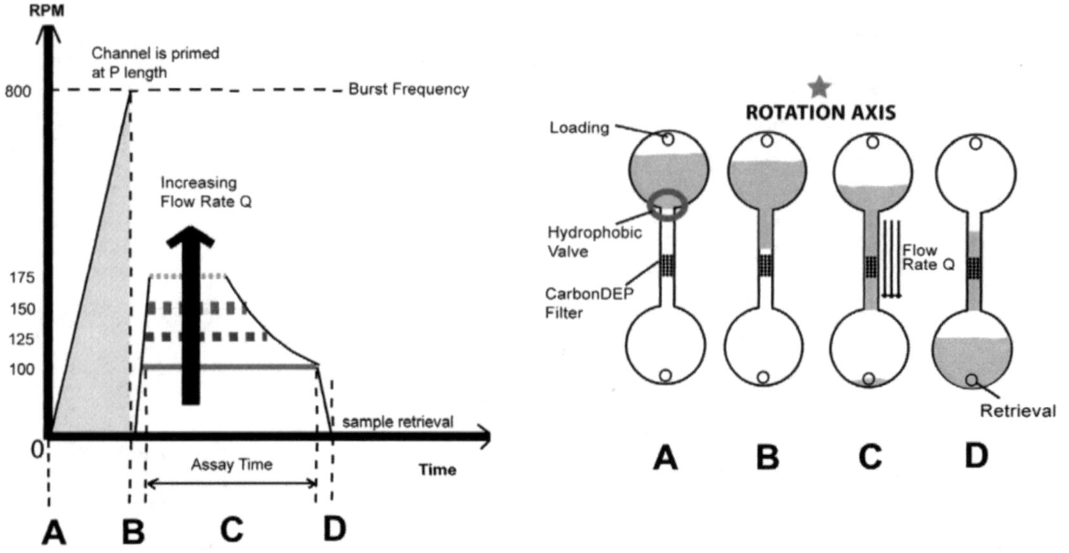


**Fig. 3** Experimental protocol illustrating the changes of spin speed during the time of the experiment and their correspondent effect in the channel. Sample is loaded at point A when CD platform is at rest. Electrodes are polarized immediately after sample loading. Hydrophobic valve is opened and channel is primed by a rapid acceleration of the platform to the burst frequency (denoted by the steep ramp from A to B). Different flow rates can be induced in the channel by the change on spin rate as shown in segment C (flow rate increases proportional to spin speed). CD is stopped to retrieve the sample at point D and electrodes are de-polarized immediately after

Filter efficiency, calculated as the ratio of the normalized number of yeast cells retrieved from the DEP chip to the normalized number of yeast cells introduced in it, is shown in Fig. 4a (normalization is done to the total number of particles, including yeast cells and latex particles, introduced or retrieved from the chip). The use of 70 μm high electrodes yields 100% filter efficiency at flow rates below 5 μl/min but the efficiency decreases as flow rate increases. The flow rate induced in the micro channel by spinning the CD platform at different speeds was also characterized and is shown in Fig. 4b.

## Discussion

The demonstrated CD-based platform offers key advantages when compared to more traditional DEP platforms where peristaltic or syringe pumps are used: 1) Footprint reduction; the demonstrated platform has the footprint of a traditional data CD and a low height profile. 2) Cost reduction; external fluidic interconnects such as tubing, sample loading docks/ports, valves, splitters and/or syringe connectors are eliminated. Furthermore, because the flow rates induced in a channel by centrifugal forces (which can be as low as 5 nl/s (45)) depend mostly on the channel dimensions and its relative position and orientation to the center of rotation, high precision motors as those required in syringe pumps are not needed in centrifugal pumping. 3) Centrifugal pumping is "contact free" and does not require any mechanical interconnects to induce a pressure gradient for pumping. The virtual nature of the pumping force results in a vented system that is immune to the presence of air bubbles. 4) The demonstrated platform is a self-contained platform. The complete fluidic network (and indeed the experiment or clinical test) can be contained in a single device (which can be disposable), be it a chip that can be embedded in a CD or the whole disk itself. 5) The demonstrated DEP platform has a modular setup and is amenable to integration with other sample preparation and detection architectures already demonstrated in CD-based platforms.

Based on our previous experience with syringe pumps (43, 46-48), we can say that both the time to run a given DEP experiment and the probability of sample leakage from the experimental setup are significantly reduced.

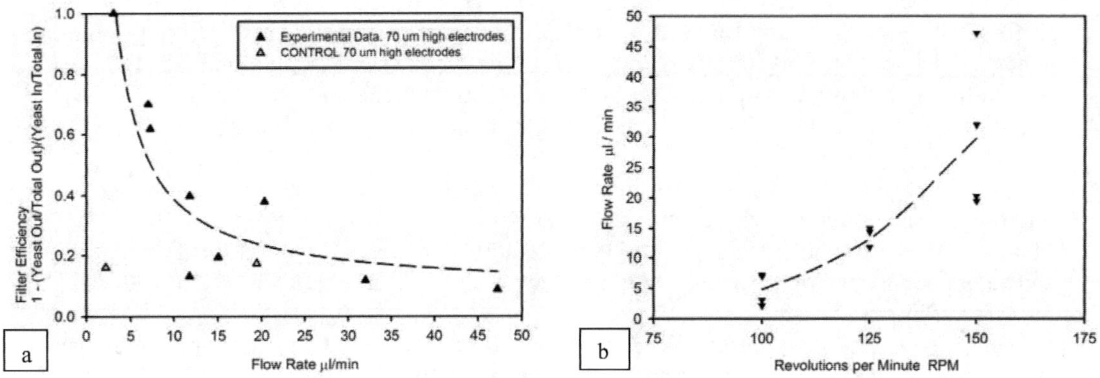


**Fig. 4** a) Filter efficiency as flow rate in the channel increases and b) flow rate in the channel as spinning speed increases

## Conclusion

The demonstrated platform enables different electrokinetic-based assays that can be conducted using a variety of different chip designs. The final goal is to add electrokinetic capabilities to other functions already demonstrated in CD platforms including valving, decanting, calibration, mixing, metering, sample splitting, and fluid separation towards an automated platform for point-of-care diagnosticss and cell therapy applications.

## References

1. Pohl,H. A. (1978), "Dielectrophoresis," Cambridge University Press: Cambridge.
2. Dalir,H., Yanagida,Y. and Hatsuzawa,T. (2009), "Probing DNA mechanical characteristics by dielectrophoresis," *Sensors Actuators B: Chem.,* 136, pp. 472-478.
3. Bakewell,D. J. and Morgan,H. (2006), "Dielectrophoresis of DNA: time-and frecuency-dependent collections on microelectrodes," *IEEE Transactions on Nanobioscience,* 5, pp. 1-8.
4. Neldecu,S. and Watson,J. H. P. (2004), "Size separation of DNA molecules by pulsed electric field dielectrophoresis," *J.Phys.D,* 37, pp. 2197-2204.
5. Asbury,C. L., Diercks,A. H. and van den Engh,G. (2002), "Trapping of DNA by dielectrophoresis," *Electrophoresis,* 23, pp. 2658-2666.
6. Chou,C. F., Tegenfeldt,J. O., Bakajin,O., et al. (2002), "Electrodeless dielectrophoresis of single- and double-stranded DNA," *Biophys.J.,* 83, pp. 2170-2179.
7. Washizu,M., Kurosawa,O., Arai,I., Suzuki,S. and Shimamoto,N. (1995), "Applications of electrostatic stretch-and-positioning of DNA," *IEEE Trans.Ind.Appl.,* 31, pp. 447-456.
8. Sung,K. E. and Burns,M. A. (2006), "Optimization of Dielectrophoretic DNA Stretching in Microfabricated Devices," *Anal.Chem.,* 78, pp. 2939-2947.

9. Kobayashi,T. and Washizu,M. (2005), "Stretch-and-positioning of single stranded DNA as a template for molecular construction," *in Proc MEMS 2005:* pp. 662-665.
10. Tegenfeldt,J. O., Prinz,C., Cao,H., et al. (2004), "Micro- and Nanofluidics for DNA Analysis," *Analytical and Bioanalytical Chemistry,* 378, pp. 1678-1692.
11. Lapizco-Encinas,B. H., Ozuna-Chacón,S. and Rito-Palomares,M. (2008), "Protein Manipulation with Insulator-Based Dielectrophoresis and DC Electric Fields," *Journal of Chromatography A,* 1206, pp. 45-51.
12. Zheng,L. F., Brody,J. P. and Burke,P. J. (2004), "Electronic manipulation of DNA, proteins, and nanoparticles for potential circuit assembly," *Biosens.Bioelectron.,* 20, pp. 606-619.
13. Suzuki,M., Yasukawa,T., Shiku,H. and Matsue,T. (2005), "Separation of live and dead microorganisms in a micro-fluidic device by dielectrophoresis," *Bunseki Kagaku,* 54, pp. 1189-1195.
14. Betts,W. B. (1995), "The potential of dielectrophoresis for the real-time detection of microorganisms in foods," *Trends in Food Science and Technology,* 6, pp. 51-58.
15. Markx,G. H., Huang,Y., Zhou,X. F. and Pethig,R. (1994), "Dielectrophoretic characterization and separation of microorganisms," *Microbiology-UK,* 140TY - JOUR Copyright 2003 SciSearch Plus, pp. 585-591.
16. Archer,G. P., Render,M. C., Betts,W. B. and Sancho,M. (1993), "Dielectrophoretic concentration of microorganisms using grid electrodes," *Microbios,* 76, pp. 237-244.
17. Gascoyne,P., Mahidol,C., Ruchirawat,M., Satayavivad,J., Watcharasit,P. and Becker,F. F. (2002), "Microsample preparation by dielectrophoresis: Isolation of malaria," *Lab on a Chip - Miniaturisation for Chemistry and Biology,* 2, pp. 70-75.
18. Sanchis,A., Brown,A. P., Sancho,M., et al. (2007), "Dielectric characterization of bacterial cells using dielectrophoresis," *Bioelectromagnetics,* 28, pp. 393-401.
19. Lapizco-Encinas,B. H., Simmons,B. A., Cummings,E. B. and Fintschenko,Y. (2004), "Insulator-Based Dielectrophoresis for the Selective Concentration and Separation of Live Bacteria in Water," *Electrophoresis,* 25, pp. 1695-1704.
20. Suehiro,J., Hamada,R., Noutomi,D., Shutou,M. and Hara,M. (2003), "Selective detection of viable bacteria using dielectrophoretic impedance measurement method," *J.Electrostatics,* 57, pp. 157-168.
21. Brown,A. P., Betts,W. B., Harrison,A. B. and O'Neill,J. G. (1999), "Evaluation of a dielectrophoretic bacterial counting technique," *Biosens.Bioelectron.,* 14, pp. 341.
22. Yang,L. (2008), "Electrical impedance spectroscopy for detection of bacterial cells in suspensions using interdigitated microelectrodes," *Talanta,* 74, pp. 1621-1629.
23. Grom,F., Kentsch,J., Müller,T., Schnelle,T. and Stelzle,M. (2006), "Accumulation and trapping of hepatitis A virus particles by electrohydrodynamic flow and dielectrophoresis," *Electrophoresis,* 27, pp. 1386-1393.
24. Lapizco-Encinas,B. H., Simmons,B. A., Cummings,E. B. and Fintschenko,Y. (2003), "High-Throughput Electrodeless Dielectrophoresis of Viruses in Polymeric Microdevices,"pp. 607-610.
25. Hughes,M. P., Morgan,H. and Rixon,F. J. (2002), "Measuring the dielectric properties of herpes simplex virus type 1 virions with dielectrophoresis," *Biochimica et Biophysica Acta-General Subjects,* 1571, pp. 1-8.
26. Ermolina,I., Milner,J. and Morgan,H. (2006), "Dielectrophoretic investigation of plant virus particles: Cow Pea Mosaic Virus and Tobacco Mosaic Virus," *Electrophoresis,* 27, pp. 3939-3948.
27. Gascoyne,P. R. C., Wang,X. B., Huang,Y. and Becker,F. F. (1997), "Dielectrophoretic separation of cancer cells from blood," *IEEE Trans.Ind.Appl.,* 33, pp. 670-678.

28. Becker,F. F., Wang,X. -., Huang,Y., Pethig,R., Vykoukal,J. and Gascoyne,P. R. C. (1995), "Separation of human breast cancer cells from blood by differential dielectric affinity," *Proc.Natl.Acad.Sci.U.S.A.,* 92, pp. 860-864.
29. Yang,J., Huang,Y., Wang,X. B., Becker,F. F. and Gascoyne,P. R. C. (1999), "Cell separation on microfabricated electrodes using dielectrophoretic/gravitational field flow fractionation," *Anal.Chem.,* 71, pp. 911-918.
30. Flanagan,L. A., Lu,J., Wang,L., et al. (2008), "Unique Dielectric Properties Distinguishing Stem Cells and Their Differentiated Progeny," *Stem Cells,* 26, pp. 656-665.
31. Stephens,M., Talary,M. S., Pethig,R., Burnett,A. K. and Mills,K. I. (1996), "The dielectrophoresis enrichment of CD34+ cells from peripheral blood stem cells harvests," *Bone Marrow Transplantation,* 18, pp. 777-782.
32. Talary,M. S., Mills,K. I., Hoy,T., Burnett,A. K. and Pethig,R. (1995), "Dielectrophoretic separation and enrichment of CD34+ cell subpopulation from bone marrow and peripheral blood stem cells," *Medical and Biological Engineering and Computing,* 33, pp. 235-237.
33. Gascoyne,P. R. C. and Vykoukal,J. V. (2004), "Dielectrophoresis-based sample handling in general-purpose programmable diagnostic instruments," *Proc IEEE,* 92, pp. 22-42.
34. Madou,Marc and Kellog,Gregory J. (1998), "LabCD: a centrifuge-based microfluidic platform for diagnostics," *Proc. SPIE,* 3259, pp. 80-93.
35. Madou,Marc, Zoval,Jim, Jia,Guangyao, Kim,Jitae and Kim,Nahui. (2006), "Lab on a CD," *Annu Rev Biomed Eng,* 8, pp. 601-628.
36. Zoval,Jim V. and Madou,Marc J. (2004), "Centrigue Based Fluidic Platforms," *Proc IEEE,* 92, pp. 140-153.
37. Wang,G-J, Hsu,W-H, Chang Y-Z and Yang,H. (2004), "Centrifugal and Electric Field Forces Dual-Pumping CD-like Microfluidic Platform for Biomedical Separation," *Biomed Microdevices,* 6, pp. 47-53.
38. Fuhr,G., Schnelle,T., Hagedorn,R. and Miller,T. (2004), "Electrode arrangement for the dielectrophoretic diversion of particles," 9/720,222,
39. Fuhr,G. and Hagedorn,R. (2002), "Method and device for manipulating particles in microsystems," 9/720,219,
40. Boettcher,M., Jaeger,M., Riegger,L., Ducrée,J., Zengerle,R. and Duschl,C. (2006), "Lab-on-chip-based cell separation by combining dielectrophoresis and centrifugation," *Biophysical Reviews and Letters,* 1, pp. 443-451.
41. Wang,Chunlei, Jia,Guangyao, Taherabadi,Lili H. and Madou,Marc J. (2005), "A Novel Method for the Fabrication of High-Aspect Ratio C-MEMS Structures," *J of MEMS,* 14, pp. 348-358.
42. Martinez-Duarte,Rodrigo and Madou,Marc J. (2009), "A novel approach to Dielectrophoresis using Carbon electrodes," *in Proceedings of the 2009 Annual Meeting of the American Electrophoresis Society:* Nashville, TN, Nov. 9-12,
43. Martinez-Duarte,R., Rouabah,H. A., Green,N. G., Madou,M. and Morgan,H. (2007), "Higher Efficiency and Throughput in Particle Separation with 3D Dielectrophoresis with C-MEMS," *in Proceedings of uTAS 2007:* Paris, France, Oct. 7-11, pp. 826-828.
44. Martinez-Duarte,R., Gorkin,R., Abi-Samra,K. and Madou,M. J. (2009), "The Integration of 3D Carbon Dielectrophoresis on a Rotating Platform," *in Proceedings of Transducers 2009:* Denver, CO, June 21-25, pp. 2147-2150.
45. Duffy,David C., Gillis,Heather L., Lin,Joe, Sheppard Jr.,Norman F. and Kellogg,Gregory J. (1999), "Microfabricated Centrifugal Microfluidic Systems: Characterization and Multiple Enzymatic Assays," *Anal Chem,* 71, pp. 4669-4678.

46. Martinez-Duarte,R., Cito,S., Collado-Arredondo,E., Martinez,S. O. and Madou,M. J. (2008), "Fluido-dynamic and Electromagnetic Characterization of 3D Carbon Dielectrophoresis with Finite Element Analysis," *Sens and Transd J,* 3, pp. 25-36.
47. Martinez-Duarte,R., Cito,S., Collado-Arredondo,E., Martinez,S. O. and Madou,M. (2008), "Fluido-dynamic and Electromagnetic Characterization of 3D Carbon Dielectrophoresis with Finite Element Analysis," *in Proceedings of NSTI Nanotech 2008:* Boston, MA, pp. 265-268.
48. Martinez-Duarte,R., Andrade-Roman,J., Martinez,S. O. and Madou,Marc J. (2008), "A High Throughput Multi-stage, Multi-frequency Filter and Separation Device based on Carbon Dielectrophoresis," *in Proceedings of NSTI Nanotech 2008:* Boston, MA, June 1-5, pp. 316-319.

# Integrated Dielectrophoretic Chip for Diagnosing Microbial Infections of Blood

**Robert S. Kuczenski**, Department of Chemical and Biomolecular Engineering, University of Notre Dame, Notre Dame, IN

Gulnaz Stybayeva, Department of Biomedical Engineering, University of California at Davis, Davis, CA

Alexander Revzin, Department of Biomedical Engineering, University of California at Davis, Davis, CA

Hsueh - Chia Chang, Department of Chemical and Biomolecular Engineering, University of Notre Dame, Notre Dame, IN

MEMS offer many advantages over traditional diagnostic technologies. For the diagnosis of sepsis, or a microbial infection of the blood, MEMS may reduce or eliminate the time-consuming microbe culturing requiring days to weeks in traditional approaches. The challenge posed by sepsis diagnosis is the relatively small microbe concentration, tens to thousands per milliliter of blood containing several millions of various types of blood cells. To overcome this challenge, we propose a three stage integrated device for microbe sorting, trapping, and identification. Using dielectrophoretic forces, the microbes are sorted from blood cells and preconcentrated on a trap facilitated by surface-bound antibodies. After preconcentration, the trapped microbes are then lysed to produce a concentrated analyte for elucidating an electrical impedance signal in the subsequent genomic identification module. The necessary sample pretreatment and device sensitivity are explored. This integrated device illustrates the potential power of leveraging MEMS technology for disease diagnosis.

See more of this Session: Biomems and Microfluidics: Biomedical Diagnostics II
See more of this Group/Topical: Topical 3: 2009 Annual Meeting of the American Electrophoresis Society (AES)

# Dielectrophoretic Responses of Human Erythrocytes

**Kaela M. Leonard**, Dave C. Swalm School of Chemical Engineering, Mississippi State University, Mississippi State, MS

J. Eric Rutan, Dave C. Swalm School of Chemical Engineering, Mississippi State University, Mississippi State, MS

Adrienne R. Minerick, Dave C. Swalm School of Chemical Engineering, Mississippi State University, Mississippi State, MS

Medical microdevices, which use small sample sizes for potentially rapid diagnosis, are currently being developed to determine human blood type based on the erythrocyte response to an alternating current dielectrophoretic field. The blood typing system of interest is the ABO-Rhesus factor system, which consists of eight blood types (A+, B+, AB+, O+, A-, B-, AB- and O-) differing by the terminal groups of the polysaccharides bound to the blood cell membrane. This work discusses the response of ABO human erythrocytes to an externally applied AC dielectrophoretic field in a custom-fabricated microdevice at 1kHz and 1MHz frequencies. At the 1kHz frequency it is observed that the erythrocytes translate very little in the field, but swell and eventually rupture. This phenomena was quantified by manually tabulating the number of unruptured cells remaining in still frame images taken from a Zeiss Axiovert inverted video microscope. The following dependencies were tested: field intensity, frequency, blood type, donor, and microdevice. Variation in rupturing rate between many of the blood types tested demonstrated that B+ blood ruptures significantly more cells in 15 minutes than any other blood type. Further, the addition of the Rhesus factor to any given ABO blood type caused a decrease in the percentage of cells ruptured over the experimental period. At the 1 MHz frequency, a pearl chain effect was observed for the majority of the blood types. This phenomenon was quantified with image intensity analysis and cell particle tracking. These results provide evidence that the membrane polysaccharides contribute to cell polarizability in a dielectrophoretic field suggesting that this electrokinetic effect may be harnessable for differentiation between blood types in the ABO typing system.

See more of this Session: Biomems and Microfluidics: Biomedical Diagnostics II
See more of this Group/Topical: Topical 3: 2009 Annual Meeting of the American Electrophoresis Society (AES)

# Nucleocytoplasmic trafficking of NF-κB studied by microfluidic electroporative flow cytometry

Jun Wang, Bei Fei, Robert L. Geahlen and Chang Lu

*Departments of Agricultural and Biological Engineering, Department of Medicinal Chemistry and Molecular Pharmacology, Weldon School of Biomedical Engineering, School of Chemical Engineering, Purdue University, West Lafayette, Indiana 47907, USA.*

## I. INTRODUCTION

Signaling pathways of a cell play a fundamental role in response to environmental change and regulate gene expression and transcription. Among several well-characterized signaling pathways, the nuclear factor-κB (NF- κB) pathway has been unique in its rapidity of activation, unusual mechanism of regulation and wide existence in all animal cell types[1, 2]. NF- κB activation represents a paradigm for regulating cell division, apoptosis and inflammation[3-6]. NF- κB is sequestered in the cytoplasm of resting cells by binding to IκB proteins which mask the nuclear localization sequence (NLS) of NF- κB[7, 8]. Upon stimulation by tumor necrosis factor-α (TNF-α) or interleukin-1 β (IL-1 β), IκB will be phosphorylated and dissociated from NF- κB by a ubiquitin-linked proteasome degradation pathway, resulting in NF- κB translocation into nucleus and induction of target gene transcription[9-11].

The translocation of activated NF- κB within cells upon stimulation is a dynamic process spatially and temporally that involves multiple kinases and nuclear transport components. To establish NF- κB translocation kinetics, traditionally electrophoretic mobility shift assay (EMSA) and western blotting are implemented to quantify NF- κB distribution between cytoplasm and nucleus[8]. However, they only provide average information of a cell population where the efficiency is largely undermined if a heterogeneous cell population is involved and its subpopulations are interested[12]. Besides, the whole process is labor intensive and requires large scale cell culture (> $10^7$ cells), precluding its use with primary cells[13]. Fluorescence or luminescence imaging would offer NF- κB localization in each cell and has shown that the translocation status is asynchronous from cell to cell[11, 14]. The small sample size (< 10) still raises the question whether those cells are representative of the entire population when constructing a model to simulate the signaling process. To improve sample size, laser scanning cytometry (LSC) has been introduced to quantify nucleus/cytoplasm distribution of intracellular proteins[15]. It has to require cells adherent to slides and plates in order to have large cytoplasmic to nuclear area ratio to facilitate spatial intensity quantification with vulnerable algorithm[16]. Traditional flow cytometry has been widely applied for single cell study with extraordinarily high throughput. Because traditional flow cytometry is intrinsically insensitive to subcellular location of proteins, it would not be useful for translocation detection and kinetic study unless being combined with cell fractionation and nuclei extraction[12, 17]. Thus, rapid, simple and robust high-throughput methods for signaling protein translocation study at single cell level are highly demanded in kinetic research and clinical application.

Any exchange of molecules, including nuclear proteins and cellular RNAs, between nucleus and cytoplasm must require transport through nuclear envelope (NE). The nuclear pore complex (NPC) allows small molecules with size limit of 20 - 40 kDa to diffuse passively through NE, whereas larger proteins require association with soluble transport receptors importin-α and importin-β[18]. With the assistance of regulating switch, RanGTP, macromolecules can be translocated in a facilitated way against a gradient of chemical activity[19]. Although many researches on import and export of proteins through NE has been done, most kinetic studies of nucleocytoplasmic transport have been performed *in vitro* and focused only on translocation of receptor-cargo complex through NPC[20]. And the kinetics and energetics of the transport remain unclear[21]. Several previous researches show that the translocation is an essentially passive process of facilitated diffusion, which suggests a physically forbidden device that reduces entropy without cost of energy[22-25].

Recently we have demonstrated the integration of electroporation and microfluidic flow cytometry for single cell study, which is referred as microfluidic electroporative flow cytometry (EFC)[26, 27]. In this report, we present the application of microfluidic EFC for NF-κB nucleocytoplasmic translocation detection and its kinetic study.

## II. EXPERIMENT

**1. Microchip fabrication**. Microfluidic EFC devices were fabricated based on PDMS using standard soft lithography method described before [28, 29]. The microscale patterns were first created using a computer-aided design software (FreeHand MX, Macromedia) and then printed out on high-resolution (5080 dpi) transparencies. The transparencies were used as photomasks in photolithography on a negative photoresist (SU-8 2010, MicroChem Corp.). The thickness of the photoresist and hence the depth of the channels was around 25 μm (measured by a Sloan Dektak3 ST profilometer). The pattern of channels in the photomask was replicated in SU-8 after exposure and development. The microfluidic channels were molded by casting a layer (~5 mm) of PDMS prepolymer mixture (General Electric Silicones RTV 615, MG chemicals) with a mass ratio of A:B = 10:1 on the photoresist/silicon wafer master treated with tridecafluoro-1,1,2,2-tetrahydrooctyl-1-trichlorosilane (United Chemical Technologies). The prepolymer mixture was cured at 85°C for 2 hours in an oven and then peeled off from the master. Glass slides were cleaned in a basic solution ($H_2O$: $NH_4OH$ (27%) : $H_2O_2$ (30%) = 5:1:1, volumetric ratio) at 75°C for an hour and then rinsed with DI water and blown dry. The surfaces of PDMS chip and a glass slide were oxidized using a plasma cleaner (Harrick). The PDMS chip was then immediately brought into contact against the slide after oxidation to form closed channels.

**2. Cell sample preparation.** The CHO/ GFP-NFκBp65 cell line (Panomics) was obtained by co-transfection of an expression vector for a fusion protein of turboGFP (Evrogen) and human NFκBp65, as well as pHyg into Chinese hamster ovary (CHO) cells. CHO cells were subcultured for every 2 days in Hams F12K media supplemented with 10% fetal bovine serum, 100 IU/ml penicillin G, 100 μg/ml streptomycin and 100 μg/ml Hygromycin. The harvested cells were either cultured in petri dish or 96 well plate for 24 hr before starvation for 4 hr in serum free media. To stimulate adherent cells, the complete Hams F12K media supplemented with IL-1 β at 20 ng/ml was used to replace starvation medium and allow cells incubated in a humidified incubator at 37℃ with 5% $CO_2$. Cells can be detached by addition of 0.25% Trypsin for 1 min and washed twice by electroporation buffer (1 mM MgSO4, 8mM $Na_2HPO_4$, 2 mM $KH_2PO_4$, and 250 mM sucrose, pH 7.2). To stimulate suspending cells, cells at a concentration of $10^6$/ml were suspended in an electroporation buffer with 20 ng/ml IL-1 β and incubated at certain temperature for designated period.

All cell samples were centrifuged at 300 x *g* for 10 min before microfluidic EFC experiment and resuspended in the electroporation buffer described with a final cell density of $10^7$ cells /ml.

**3. Fluorescence microscopy.** The microfluidic device was mounted on an inverted fluorescence microscope (IX-71, Olympus, Melville, NY) with a 40X dry objective (NA= 0.60). The epifluorescence excitation was provided by a 100W mercury lamp, together with brightfield illumination. The excitation and emission from p65-GFP expressing cells were filtered by a fluorescence filter cube (exciter HQ480/40, emitter HQ535/50, and beamsplitter Q505lp, Chroma technology Corp.). To observe cells after electroporation, cells were transferred from the microchip reservoir to a 96 well plate and then centrifuged for 10 min at 300 x *g* to settle the cells to the bottom before imaging under the microscope. The real-time observation and imaging of p65-GFP translocation in suspending cells were conducted by placing cell samples on a poly-L-lysine coated glass slide mounted on the microscope, while the bulk cells were stimulated and incubated under certain temperature.

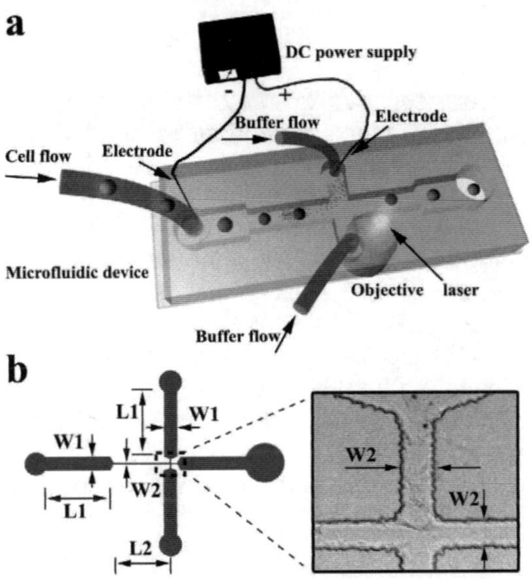


**Figure 1** (a) Schematic setup and principle of electroporative flow cytometry. The microfluidic device is mounted on a microscope that allow incident laser from a 40X objective to focus on cells in the downstream of narrow sections. The microchannel has 3 inlets, one for cell sample and the other two for buffer flow. A DC power supply connects with two electrodes in the illustrated inlets to generate high electric field in narrow sections to electroporate cells and release intracellular proteins. (b) Layout of the design for microfluidic channels. The dimensions are: $W_1$=400 µm, $L_1$= 3 mm, $W_2$= 40 µm. The length $L_2$ is designed as 4 mm, 8 mm and 12 mm which corresponds to 60 ms, 120 ms and 180 ms electroporation duration time in the text assuming the same flow rate from all inlets. (Inset) The intersection of narrow sections. Released intracellular materials accumulated in the top channel after electric field at 800 V/cm maintained for 5 min.

**4. Microchip operation and signal detection.** The setup of the apparatus is shown in **Fig. 1**. The microchip was mounted on the microscope where the center of the objective was aligned to 150 µm downstream after the intersection of two narrow sections (Fig. 1b). To prevent released proteins sticking to channel walls during electroporation, the microfluidic channel was flowed with 10% pluronic F68 solution (Sigma) for one hour followed with the electroporation buffer for 15 min prior to all experiments. The 3 inlets of the channel were connected to a syringe pump (PHD infusion pump, Harvard Apparatus) through plastic tubing. The volumetric flow rates were set at 2 µl/min for the sample channel inlet and 10 µl/min for each of two side channel inlets. With a cell density of $10^7$ cells/mL, 250-300 cells flowed through the laser detection spot per second. To prevent released proteins from electroporated cells interfering with fluorescent signal of cells, anode was placed in one of side inlets while cathode was located in the sample inlet. A high voltage power supply (PS350, Stanford Research Systems) was used to generate a constant direct current (DC) voltage in between the electrodes.

An air-cooled 100mW argon ion laser (Spectra-Physics) at 488 nm was applied as the excitation source for laser-induced fluorescence. The laser beam was spectrally filtered by a 10LF10-488 bandpass filter (Newport Corp.) before its intensity was adjusted by neutral density filters (Newport Corp.). The laser was introduced into the microscope through laser port B (Olympus) and a fluorescence filter cube (505DCLP dichroic beamsplitter, D535/40 emission filter, Chroma Technology Corp.) before it was finally focused by the objective into the microfluidic channel. The emission light was collected by the same objective and converted into current by a photomultiplier tube (R9220, Hamamatsu) biased at 500 V. The photocurrent was amplified by a low noise current preamplifier (SR570, Standard Research System) with the cutoff frequency and sensitivity set at 30K Hz and 20 μA/V, respectively. The current was then converted to voltage and input into a PCI data acquisition card (PCI-6254, National Instruments) operated by LabView software (National Instruments, Austin). The data were processed by programs written in MATLAB to extract histograms of the fluorescence from a cell population. The data were presented in 4 decades (from 0.001 to 10 V) logarithmic histograms with 256 channels. The voltage signal ranging from 1 mV to 1 V was converted to 4 decade logarithmic voltage scale and then 256 scale channels, due to the relative small sample size of 6000-10000 cells in each histogram.

**5. Western blot analysis** To employ biochemical means to visualize the distribution of RelA-GFP among different cellular compartments, cytosolic and nuclear fractionations of CHO cells were separated according to the standard protocol. Basically, CHO cells were starved in serum free F12K medium for 4 h before stimulation. IL-1β at a final concentration of 20 ng/ml was used to treat the cells. After the stimulation, cells were spun down using a table-top spinner for 30 sec and washed with 1X PBS supplemented with 200 μg/mL STI (soybean trypsin inhibitor). Cells were then suspended in 400 μL cold buffer A (10 mM Hepes, pH 7.9, 10 mM KCl, 0.1 mM EDTA, 0.1 mM EGTA, 1 mM DTT, 0.5 mM PMSF, 1 mM NaOV, 2X protease inhibitor cocktail) by gently pipetting with a yellow tip for 8 times before set on ice to swell for 10 min. 8 μL of 10% NP-40 was added to achieve a final concentration of 0.2%. The homogenates were incubated on ice for another 5 min and then spun down at 14,000 rpm for 30 sec. The supernatants were saved as cytosolic fractions while the nuclear parts were washed with 200 μL detergent-free buffer A. The nuclear parts were then resuspended in 300 μL 1X PSM by using a 1-mL syringe with a 22G1/2 needle in order to shear the chromosome DNA. Resulting samples were then subjected to SDS-PAGE and Western blot analysis with an anti-RelA antibody.

**6. Fluorescence intensity calibration**. The quantification of fluorescence intensity from single cells using Molecules of Equivalent Soluble Fluorophores (MESF) units were described before [26, 30, 31]. In brief, a standard curve was first established by screening beads with different predefined fluorescence quantities (high level Quantum FITC MESF Kits, Bangs Laborotories, Inc.) using our microfluidic EFC devices. The peak channel numbers of beads was plotted against their known MESF values. The mean percentage of GFP of a cell population, $P_{GFP}$, (by assuming that p65-GFP cell population without electroporation has a mean percentage of 100% and cell population without GFP has a mean percentage of 0) was calculated using following formulae

$$P_{GFP} = \frac{[GFP]}{[GFP]_{E=0}} \times 100\% \tag{4}$$

where [GFP] is MESF value for p65-GFP cell population under various electric field, calculated from mean fluorescence intensity, and [GFP] is MESF value for p65-GFP cell population without electric field.

## III. RESULTS AND DISCUSSION

The setup and layout of microfluidic EFC device is shown in **Fig. 1**. The microfluidic EFC device mounted on a microscope was fabricated with PDMS/glass slide by standard lithography described before[29]. There were 1 outlet for waste and 3 inlets for the sample and buffer flow. A common DC power supply was connected with two platinum electrodes penetrating through PDMS to the sample channel inlet and a side channel inlet to establish electric potential between them (**Fig. 1a**). The microfluidic channels were designed to incorporate geometry variations so that a relatively large electric field was generated in narrow sections, whereas a negligible electric field resided in wide sections. Given the uniform depth of the channels at 25 µm, the electric field intensity at wide section 1 is only 1/10 of that in narrow section 2 since the width $W_1$ is 10 times of $W_2$[32] (**Fig. 1b**). In the range of field intensities from 0 V/cm to 1000 V/cm in narrow sections, the field intensities in wide sections were too weak to compromise the cell membrane when an overall voltage within ?V was added[26]. Hence the duration time that cells flow through the narrow section with length $L_2$ determined the duration that cells were electroporated. $L_2$ was designed to be 4 mm, 8 mm and 12 mm, which corresponded to 60 ms, 120 ms and 180 ms duration time as the sample flow rate was maintained at 2 µl/min. In all the following text, the duration time and electric field intensity refer to those in narrow section of sample channel. The flow rates of side channels were set at 10 µl/min to sufficiently focus sample stream and allow cells to pass through laser detection area one by one (**Fig. 1a**). Throughput of 250~300 cells/s is obtained after data acquisition and collection by Labview and Matlab programs in our lab.

The stimulation process of CHO cell line expressing p65-GFP is shown in **Fig. 2**. p65 is a subunit of NFκB which is sequestered in cytoplasm before activation. When the cells are stimulated with 20 ng/ml IL-1 β at 37 ℃, NF κB will dissociate with inhibitor protein IκB which is degraded by IκB kinase (IKK). The freed NLS sequence of NFκB binds to Importin and drives NFκB to transport through nuclear pore complex (NPC) into nucleus with assistance of RanGTP[18]. With persistent stimulation by IL-1 β, we observed that the translocation completed rapidly within 30 min in both adherent cells and suspending cells (**Fig. 2**). Nearly 100% cells have p65-GFP redistributing into nuclei. This translocation can be prolonged for 120 min without obvious difference in terms of fluorescence intensity ratio of nucleus to cytoplasm. Thus without specific note, cells were stimulated for 30 min at 37 ℃ when translocated cells were desired in the following experiments on microfluidic

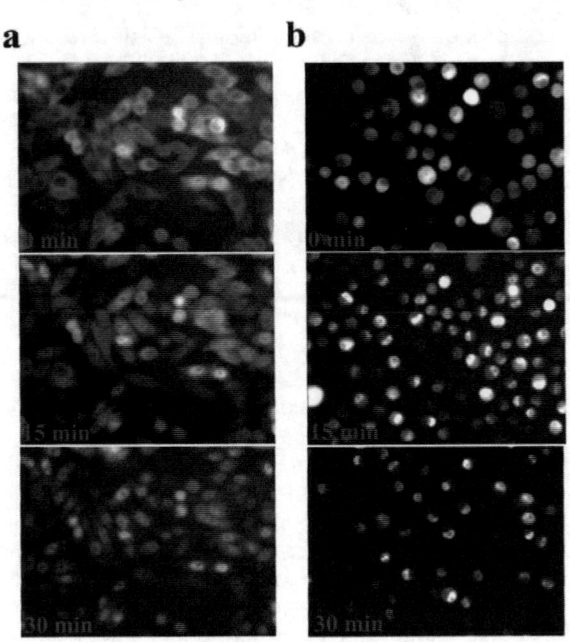


**Figure 2** Time-lapse fluorescence images of CHO cells expressing p65-GFP translocating from cytoplasm into nucleus after stimulation by 20 ng/ml IL-1 β at 37 ℃. (a) Stimulation of adherent CHO cells. (b) Stimulation of suspending CHO cells where cells were detached from plate and resuspended in phosphate buffer.

EFC.

We used microfluidic EFC to test release of intracellular proteins under different electric field intensities (**Fig. 3a**). The cells without stimulation were pumped through the sample channel where electric fields were located. The fluorescence intensities were recorded and presented in logarithmic histograms binned into 256 channel scale. The y axis with percentile frequency was used to eliminate effect of sample size on histogram comparison. The histograms have a significant shift from 0 V/cm to 800 V/cm when the duration time is set at120 ms. The difference between mean channel number of the histogram at 0 V/cm and 800 V/cm reaches 67 channels, which denotes that 91.1% p65-GFP is depleted from cells in average. It is estimated that the protein p65-GFP has a negative charge in the electroporation buffer with pH 7.4 since isoelectric point of NFκB subunits is 5.4~5.9 and GFP is 5.0[33, 34]. The released protein p65-GFP and its complex will be dispersed due to diffusion and electrophoretic motion in the narrow section. The design of anode electrode in a side channel also contributes to dispersion since the released proteins at the intersection will experience a perpendicular electrophoretic force directing to the top inlet (**Fig. 1a**). In this way we reduced interference of released fluorescent protein and cell debris. **Fig. 3b** shows the results obtained by microfluidic EFC comparing fluorescence intensity of cells with stimulation and cells without stimulation under various electric field intensities at 120 ms duration. Before the application of electric field, the histograms of those two cell populations are totally overlapped. The stimulation process itself only translocates p65-GFP to different subcellular location but not change the expression level. Similar to conventional flow cytometry, microfluidic EFC without a voltage on the device is not sensitive to subcellular fluorophore distribution. Both of the histograms shift rapidly to lower fluorescence level from 500 V/cm due to electroporative release of p65-GFP. And the histogram of cells without stimulation is always in the left of that of cells with stimulation, which indicates that p65-GFP in cells without stimulation can be easier to be released outside. And the higher electric field intensity it is applied, the larger difference the two histograms attain.

The performance of microfluidic EFC is characterized by investigating the effect of electric field and duration time shown in **Fig. 3c, d and e**. The mean channel number of each histogram was extracted to represent the fluorescence intensity of

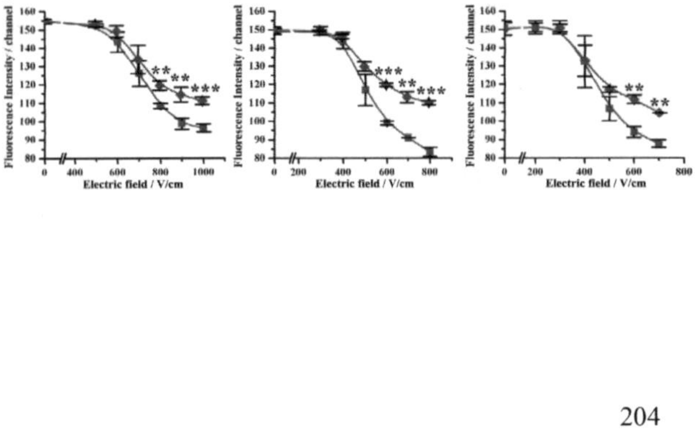


**Figure 3** (a) Shift of fluorescence intensity histograms generated by p65-GFP cells processed in a microfluidic EFC device under electric fields from 0 V/cm to 800 V/cm. The electroporation duration time is 120 ms. (b) Comparison of fluorescence intensity histograms between stimulated cells (red) and cells without stimulation (blue) under electric fields from 0 V/cm to 800 V/cm. The electroporation duration time is 120 ms. The stimulation was conducted at 37 ℃ for 30 min. (c) Variation of the mean fluorescence intensity value of the cell population at different field intensities with and without stimulation by IL-1 β. Cell populations with stimulation (red) and without stimulation (blue) are compared under field intensities from 0 V/cm to 1000 V/cm when the duration time is 60 ms. The experiments were replicated in triple times to generate error bars. Longer electroporation duration time of 120 ms and 180 ms are shown in (c) and (d), respectively. The difference between two populations with the same electric parameter is statistically significant when P value is less than 0.01 (**) and 0.001 (***).

the cell population at certain electric parameter. Essentially both cell populations released p65-GFP in the range of electric field intensity tested in **Fig. 3**. At 60 ms duration time, the release starts from 600 V/cm. At longer duration time of 120 ms and 180 ms, the release happens at much lower electric field intensity just above 400 V/cm, which has been known as the threshold of electroporation[29]. In all three duration time tested, the trend is the same that the difference between cells with stimulation and cells without stimulation increases as the electric field intensity rises. However, 120 ms duration time seems to best differentiate those two populations among 3 durations. At the highest electric field intensity of 800 V/cm, there is 31 channel difference, or ~3 times difference of absolute fluorescence intensity.

When the transmembrane potential on a cell over critical value of ~0.25 - 1 V, the cell membrane will be compromised and generate electro-pores which allow exchange of molecules inside and outside of the cell[35]. Although the electroporation process has been very well studied, the release of subcellular protein during electroporation has not been understood. Presumably proteins closer to plasma membrane are more subject to releasing when electro-pores are generated on the membrane. Convection outside of membrane due to fluidic flow would also enhance diffusion and dispersion in bulk fluid. On the contrary, proteins in the nucleus are less possible to release due to obstruction of NE. The binding of p65-GFP to chromosome would also prevent releasing of p65-GFP since such a large complex is presumably hard to pass through NE and electro-pores on plasma membrane. Thus cells with stimulation can retain more p65-GFP proteins within cells than those without stimulation under the same electric parameters.

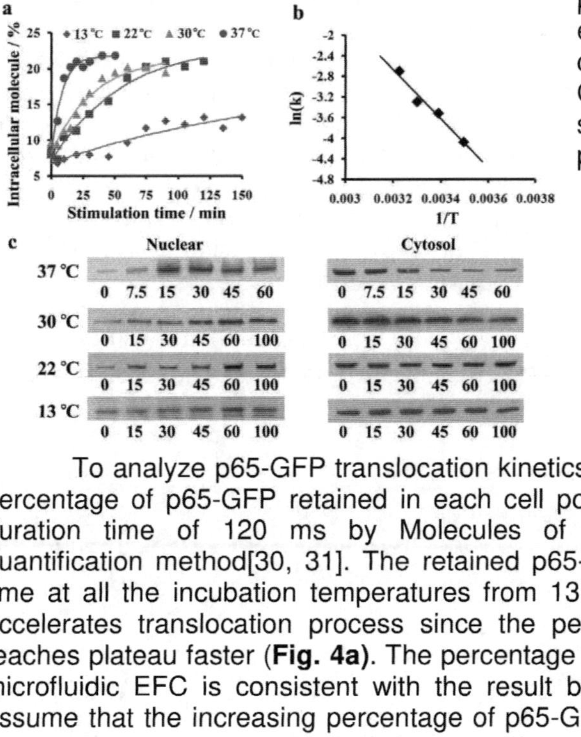


**Figure 4** Nuclear transport kinetics of NFκB. (a) Effect of temperature on nuclear import rate. The dots are experimental data which are fitted Eq. 2 shown by curves. (b) Arrhenius plot of the transport rates (dots) at different temperatures. It shows linear behavior with $R^2$=0.97. (c) Western blotting result of NFκB in nuclear and cytosol at different temperatures and stimulation times (min).

To analyze p65-GFP translocation kinetics with microfluidic EFC, we firstly quantify the percentage of p65-GFP retained in each cell population with electric field of 700 V/cm and duration time of 120 ms by Molecules of Equivalent Soluble Fluorophores (MESF) quantification method[30, 31]. The retained p65-GFP increases significantly with stimulation time at all the incubation temperatures from 13 ℃ to 37 ℃ (**Fig. 4a**). Higher temperature accelerates translocation process since the percentage of retaining intracellular p65-GFP reaches plateau faster (**Fig. 4a)**. The percentage of retained p65-GFP within cells analyzed by microfluidic EFC is consistent with the result by western blotting (**Fig. 4c**). Therefore, we assume that the increasing percentage of p65-GFP by microfluidic EFC is the amount that is transported into nucleus. With that assumption, we found the transport kinetics complied with first-order process, i.e.

$$\frac{dP}{dt} = k(P_{SS} - P) \tag{1}$$

where $P_{ss}$ is the amount of p65-GFP within cells that reaches steady state after translocation and $k$ is the first-order rate constant. Or it can be expressed as

$$P = P_{SS}[1-(1-\frac{P_0}{P_{SS}})\exp(-kt)] \quad (2)$$

where $P_0$ is the amount of p65-GFP at 0 min. The fitting results show that the transport should end up with the same amount of p65-GFP translocating into nucleus at all temperatures. We found $k$ was not linear with temperature T, which indicated that the process of transporting NFκB into nucleus was not a simple diffusion[36] (since the diffusion coefficient can not be fitted in Einstein diffusion equation). From the plot of ln($k$) with 1/T (**Fig. 4b**), the slope of the fitting line can be obtained as -4789, according to Arrhenius equation

$$\ln(k) = -\frac{E_a}{R}\frac{1}{T} + C \quad (3)$$

where $E_a$ is the activation energy, $R$ is gas constant and $C$ is a constant. In a consequence the activation energy $E_a$ of the translocation process would be 39.8 KJ/mol, or $6.61\times10^{-20}$ J for each molecule. This value of $E_a$ is similar as the activation energy required for NLS-GFP translocation in *Saccharomyces cerevisiae*[37], which indicates that interaction of NLS tail of p65-GFP with NLS receptor dominates the translocation process. Meanwhile this high $E_a$ value suggests that the translocation of p65-GFP protein is mediated and not diffusion rate limited.

## IV. CONCLUSION

A microfluidic platform is validated to detect NF-κB translocation and demonstrated to study nuclear transport kinetics *in vivo*. Under certain electric parameters, cells with NF-κB translocation into nucleus essentially release less intracellular proteins and consequently have higher fluorescence intensity than those without it. And the differences expand as electric field rises, while there is an optimal electroporation time of ~120 ms that can differentiate stimulated and unstimulated cells at largest. We also trafficked NF- κB nuclear transport *in vivo* with time under different incubation temperature by monitoring calibrated NF- κB level in whole cell with EFC. Using a first-order kinetic model, we show how the system simply simulates a biochemical reaction in the perspective of thermodynamics. The transport rate can fit in Arrhenius equation and a high activation energy is generated, which indicates that the nucleocytoplasmic translocation is an energy consumption process. We envision that microfluidic EFC will be an important tool in studying biological processes involving nucleocytoplasmic trafficking and biochemical kinetics.

## ACKNOWLEDGEMENT

This work is supported by Wallace H. Coulter Foundation, NSF grant CBET-0747105, a cooperative agreement with the ARS-USDA, project number 1935 42000-035, through the Center for Food Safety Engineering at Purdue University and Bilsland Fellowship (J.W.).

## REFERENCES

1. Karin M & Ben-Neriah Y (2000) Phosphorylation meets ubiquitination: The control of NF-kappa B activity. *Annual Review of Immunology* 18, pp. 621-+.
2. Karin M & Lin A (2002) NF-kappa B at the crossroads of life and death. *Nature Immunology* 3, pp. 221-227.
3. VanAntwerp DJ, Martin SJ, Kafri T, Green DR & Verma IM (1996) Suppression of TNF-alpha-induced apoptosis by NF-kappa B. *Science* 274, pp. 787-789.
4. Wang CY, Mayo MW & Baldwin AS (1996) TNF- and cancer therapy-induced apoptosis: Potentiation by inhibition of NF-kappa B. *Science* 274, pp. 784-787.

5. Ryan KM, Ernst MK, Rice NR & Vousden KH (2000) Role of NF-kappa B in p53-mediated programmed cell death. *Nature* 404, pp. 892-897.
6. Li QT & Verma IM (2002) NF-kappa B regulation in the immune system. *Nature Reviews Immunology* 2, pp. 725-734.
7. Malek S, Chen Y, Huxford T & Ghosh G (2001) I kappa B beta but not I kappa B alpha, functions as a classical cytoplasmic inhibitor of NF-kappa B dimers by masking both NF-kappa B nuclear localization sequences in resting cells. *Journal of Biological Chemistry* 276, pp. 45225-45235.
8. Hoffmann A, Levchenko A, Scott ML & Baltimore D (2002) The I kappa B-NF-kappa B signaling module: Temporal control and selective gene activation. *Science* 298, pp. 1241-1245.
9. Lernbecher T, Muller U & Wirth T (1993) Distinct Nf-Kappa-B Rel Transcription Factors Are Responsible for Tissue-Specific and Inducible Gene Activation. *Nature* 365, pp. 767-770.
10. Chen LF, Fischle W, Verdin E & Greene WC (2001) Duration of nuclear NF-kappa B action regulated by reversible acetylation. *Science* 293, pp. 1653-1657.
11. Nelson DE, Ihekwaba AEC, Elliott M, Johnson JR, Gibney CA, Foreman BE, Nelson G, See V, Horton CA, Spiller DG, Edwards SW, McDowell HP, Unitt JF, Sullivan E, Grimley R, Benson N, Broomhead D, Kell DB & White MRH (2004) Oscillations in NF-kappa B signaling control the dynamics of gene expression. *Science* 306, pp. 704-708.
12. Blaecke A, Delneste Y, Herbault N, Jeannin P, Bonnefoy JY, Beck A & Aubry JP (2002) Measurement of nuclear factor-kappa B translocation on lipopolysaccharide-activated human dendritic cells by confocal microscopy and flow cytometry. *Cytometry* 48, pp. 71-79.
13. Noursadeghi M, Tsang J, Haustein T, Miller RF, Chain BM & Katz DR (2008) Quantitative imaging assay for NF-kappa B nuclear translocation in primary human macrophages. *Journal of Immunological Methods* 329, pp. 194-200.
14. Nelson G, Paraoan L, Spiller DG, Wilde GJC, Browne MA, Djali PK, Unitt JF, Sullivan E, Floettmann E & White MRH (2002) Multi-parameter analysis of the kinetics of NF-kappa B signalling and transcription in single living cells. *Journal of Cell Science* 115, pp. 1137-1148.
15. Deptala A, Bedner E, Gorczyca W & Darzynkiewicz Z (1998) Activation of nuclear factor kappa B (NF-kappa B) assayed by laser scanning cytometry (LSC). *Cytometry* 33, pp. 376-382.
16. George TC, Fanning SL, Fitzgeral-Bocarsly P, Medeiros RB, Highfill S, Shimizu Y, Hall BE, Frost K, Basiji D, Ortyn WE, Morrissey PJ & Lynch DH (2006) Quantitative measurement of nuclear translocation events using similarity analysis of multispectral cellular images obtained in flow. *Journal of Immunological Methods* 311, pp. 117-129.
17. Foulds S (1997) Novel flow cytometric method for quantifying nuclear binding of the transcription factor nuclear factor kappa B in unseparated human monocytes and polymorphonuclear cells. *Cytometry* 29, pp. 182-186.
18. Riddick G & Macara IG (2005) A systems analysis of importin-alpha-beta mediated nuclear protein import. *Journal of Cell Biology* 168, pp. 1027-1038.
19. Gorlich D & Kutay U (1999) Transport between the cell nucleus and the cytoplasm. *Annual Review of Cell and Developmental Biology* 15, pp. 607-660.
20. Timney BL, Tetenbaum-Novatt J, Agate DS, Williams R, Zhang WZ, Chait BT & Rout MP (2006) Simple kinetic relationships and nonspecific competition govern nuclear import rates in vivo. *Journal of Cell Biology* 175, pp. 579-593.
21. Kopito RB & Elbaum M (2007) Reversibility in nucleocytoplasmic transport. *Proceedings of the National Academy of Sciences of the United States of America* 104, pp. 12743-12748.

22. Nakielny S & Dreyfuss G (1997) Import and export of the nuclear protein import receptor transportin by a mechanism independent of GTP hydrolysis. *Current Biology* 8, pp. 89-95.
23. Keminer O, Siebrasse JP, Zerf K & Peters R (1999) Optical recording of signal-mediated protein transport through single nuclear pore complexes. *Proceedings of the National Academy of Sciences of the United States of America* 96, pp. 11842-11847.
24. Ribbeck K, Kutay U, Paraskeva E & Gorlich D (1999) The translocation of transportin-cargo complexes through nuclear pores is independent of both Ran and energy. *Current Biology* 9, pp. 47-50.
25. Rout MP & Aitchison JD (2001) The nuclear pore complex as a transport machine. *Journal of Biological Chemistry* 276, pp. 16593-16596.
26. Wang J, Bao N, Paris LL, Wang HY, Geahlen RL & Lu C (2008) Detection of kinase translocation using microfluidic electroporative flow cytometry. *Anal Chem* 80, pp. 1087-1093.
27. Bao N, Zhan YH & Lu C (2008) Microfluidic electroporative flow cytometry for studying single-cell biomechanics. *Anal Chem* 80, pp. 7714-7719.
28. Wang HY & Lu C (2006) Electroporation of mammalian cells in a microfluidic channel with geometric variation. *Anal Chem* 78, pp. 5158-5164.
29. Wang J & Lu C (2006) Microfluidic cell fusion under continuous direct current voltage. *Applied Physics Letters* 89, -.
30. Bjellqvist B, Hughes GJ, Pasquali C, Paquet N, Ravier F, Sanchez JC, Frutiger S & Hochstrasser D (1993) The Focusing Positions of Polypeptides in Immobilized Ph Gradients Can Be Predicted from Their Amino-Acid-Sequences. *Electrophoresis* 14, pp. 1023-1031.
31. Penna TCV, Chiarini E, Machoshvili IA, Ishii M & Pessoa A (2002) Intracellular release of recombinant green fluorescent protein (gfp(uv)) from Escherichia coli. *Applied Biochemistry and Biotechnology* 98, pp. 791-802.
32. Teissie J & Rols MP (1993) An Experimental Evaluation of the Critical Potential Difference Inducing Cell-Membrane Electropermeabilization. *Biophysical Journal* 65, pp. 409-413.
33. Wang LL, Gaigalas AK, Abbasi F, Marti GE, Vogt RF & Schwartz A (2002) Quantitating fluorescence intensity from fluorophores: Practical use of MESF values. *Journal of Research of the National Institute of Standards and Technology* 107, pp. 339-353.
34. Martin BR, Giepmans BNG, Adams SR & Tsien RY (2005) Mammalian cell-based optimization of the biarsenical-binding tetracysteine motif for improved fluorescence and affinity. *Nature Biotechnology* 23, pp. 1308-1314.
35. Ribbeck K & Gorlich D (2001) Kinetic analysis of translocation through nuclear pore complexes. *Embo Journal* 20, pp. 1320-1330.
36. Shulga N, Roberts P, Gu ZY, Spitz L, Tabb MM, Nomura M & Goldfarb DS (1996) In vivo nuclear transport kinetics in Saccharomyces cerevisiae: A role for heat shock protein 70 during targeting and translocation. *Journal of Cell Biology* 135, pp. 329-339.
37. Duffy DC, McDonald JC, Schueller OJA & Whitesides GM (1998) Rapid prototyping of microfluidic systems in poly(dimethylsiloxane). *Analytical Chemistry* 70, pp. 4974-4984.

# A Microfluidic System for Fiber Fluorescence in Situ Hybridization

Xiaozhu Wang, Chemical and Biomedical Engineering, Florida State University, Tallahassee, FL

Peipei Zhang, Chemical and Biomedical Engineering, Florida State University, Tallahassee, FL

**Jingjiao Guan**, Chemical and Biomedical Engineering, Florida State University, Tallahassee, FL

Fluorescence in situ hybridization (FISH) is a technique that relies on complimentary binding between fluorescently labeled nucleic acid probes and genomic DNA for detecting specific DNA sequences. It is widely used for both fundamental biological research and medical diagnosis. Conventional FISH is performed at the level of single cells, but it is limited by low resolution due to highly compacted genome in the nuclei. Fiber FISH is an advanced FISH method based on stretching free DNA extracted from cells onto a solid surface. It possesses higher resolution, but suffers from breakage of long genomic DNA and loss of identity of DNA related to the cells. We introduce a microfluidic system able to extract genomic DNA from single cells. While the DNA can be well stretched by precise control of flow conditions, the relationship between the DNA and cells is also retained. High resolution FISH is performed on the stretched DNA in the microfluidic system for cancer diagnosis.

See more of this Session: Biomems and Microfluidics: Biomedical Diagnostics II
See more of this Group/Topical: Topical 3: 2009 Annual Meeting of the American Electrophoresis Society (AES)

# Microfluidics Meets Bioenergy: Challenges and Opportunities

**Rajiv Bharadwaj**, The Joint BioEnergy Institute, Emeryville, CA

Anup K. Singh, Biosystems Research and Development, Sandia National Laboratories, Livermore, CA

There is a growing interest in the production of carbon-neutral biofuels from cellulosic biomass. The key steps include feedstocks development, biomass deconstruction, and fuels synthesis. Each of these steps require a suite of analytical techniques to enable optimization of process development. Microfluidics-enabled analytical tools can have a significant impact in the field of bioenergy. This talk will provide an overview of microfluidics research at the Joint BioEnergy Institute. We will describe the challenges of the biomass-to-fuels process and the opportunities for microfluidic solutions for biomass analysis analysis, cellulase optimization, and plant cell wall engineering.

See more of this Session: Biomems and Microfluidics: Novel Applications
See more of this Group/Topical: Topical 3: 2009 Annual Meeting of the American Electrophoresis Society (AES)

# Microfluidics Enabled Parallel Co-Cultivation of Compartmentalized Human Microbiota: Towards Elucidation of Microbial Interactions

Jihyang Park, Chemical Engineering, University of Michigan, Ann Arbor, MI

Mark A. Burns, Chemical Engineering, University of Michigan, Ann Arbor, MI

Xiaoxia Nina Lin, Department of Chemical Engineering, Department of Biomedical Engineering, University of Michigan, Ann Arbor, MI

In nature, diverse microbial communities exist with specific cell population profiles, which affect and adapt to various environments. Among them, the human body is a representative host for microbial communities, such as oral, skin, vaginal and gut microbiota. The genes from these microbial populations greatly outnumber human genes, contribute to human metabolism, and increases genetic complexity in human. Despite their relevance, the human microbiota remain largely unknown, due to their complexity and uncultivability. Most of the previous work were based on genetic identification, while microbial intercellular networks have been hardly investigated so far.

In order to elucidate microbial interactions in human microbiota, we are constructing microfluidic devices capable of compartmentalizing subsets of the total human microbiota and cultivating them. Since the gut microbiota habours the largest population among the human microbiota, we used the fecal sample collected from gnotobiotic mice colonized with human gut microbiota. The sample was diluted and then compartmentalized in multiple droplets, which were serially generated in Parylene-coated glass devices adapting slanted T-junction geometry. Localization of each subset, in microdroplets (< 10nl) dispersed in immiscible oil, separates essential intercellular links for growth from the whole complicated network. Depending on the initial cell concentration in the droplets, different numbers of cell clusters comprising multiple species were developed in each droplet. Droplets carrying fully-grown microbial communities were retrieved from the device using microfluidic control, after their growths were monitored under the microscope. Off-chip PCR of 16S rRNA genes with universal bacterial primers from the retrieved culture shows the species profiles in subsets of the microbiota. Based on these result, we can start to elucidate the intercellular relationship in the community. This parallelized microfluidic compartmentalization and on-chip cultivation platform can also be extended for the investigation of microbial communities in other environments such as the human oral microbiota.

See more of this Session: Biomems and Microfluidics: Novel Applications
See more of this Group/Topical: Topical 3: 2009 Annual Meeting of the American Electrophoresis Society (AES)

# A Microfluidic Migration Assay for Single-Cell Tracking in Well-Controlled and Non-Flowing Gradient Fields

**Edward S. Park**, Chemical and Biomolecular Engineering, Georgia Institute of Technology, Atlanta, GA

Hang Lu, Chemical and Biomolecular Engineering, Georgia Institute of Technology, Atlanta, GA

The investigation of cellular response to soluble gradients is important for understanding migration phenomena, such wound healing, embryonic development, and cancer metastasis. Traditional migration assays provide end-point (e.g. Boyden, transwell) or time-lapse (e.g. Dunn) information; however, these methods employ gradients with suboptimal control and stability. Microfluidic assays have emerged, due in large part to the stability offered by laminar flows. Most microfluidic methods require either direct flow over cells or diffusion in static fields, which could lead to undesired shear effects or lack of control, respectively. Such trade-offs pose challenges to achieving optimal microenvironments in migration assays.

We address the abovementioned challenges by developing a device that produces a soluble concentration gradient that is (i) fully stable, (ii) switchable, and (iii) generates zero net flow through a cell culture chamber. The device contains a novel microfluidic architecture, where a flowing sheet acts as a gradient "source." Gradient information from the source is translated into the chamber by diffusion with high fidelity. As it is sealed in all other directions, there is no net flow through the chamber, which minimizes the possibility of shear-induced cellular behavior. An upstream flow network selects streams for different stages of each experiment, such as cell loading/seeding, growth, serum deprivation, or gradient activation. Finite element models show a steady-state gradient with trace velocities, and the gradient is verified via intensity measurements of fluorescent markers. We validated the design by measuring responses of NIH/3T3 cells to growth factor gradients. Cells are seeded at low density in chambers (1000x300 μ m) pre-incubated with extracellular matrix (i.e. fibronectin). Single cells are tracked by calculating centroid trajectories from traces of cell edges, yielding step-velocities and morphological readouts (e.g. aspect ratio, angular orientation). These data are compared by diverse graphical and statistical means, establishing the effectiveness of the device in performing high-resolution migrational studies.

See more of this Session: Biomems and Microfluidics: Novel Applications
See more of this Group/Topical: Topical 3: 2009 Annual Meeting of the American Electrophoresis Society (AES)

# Total Internal Reflection Fluorescence Flow Cytometry

*Jun Wang, Ning Bao, Leela L. Paris, Robert L. Geahlen and Chang Lu*[a)]

*Departments of Agricultural and Biological Engineering, Department of Medicinal Chemistry and Molecular Pharmacology, School of Chemical Engineering, Birck Nanotechnology Center and Bindley Bioscience Center, Purdue University, West Lafayette, Indiana 47907, USA.*

## I. INTRODUCTION

The cell membrane is the location where most biological signals are sent or received by a cell. Key events such as organelle and protein trafficking take place close to the cell surface and the visualization of these processes has been largely relying on imaging tools such as Total Internal Reflection Fluorescence Microscopy (TIRFM) [1-7]. In a typical TIRFM setup, evanescent waves that are generated by totally reflected incident light at a glass-water interface penetrate into an adherent cell on the glass surface with a depth less than 200 nm. Only the features and events at the plasma membrane and the cytoplasmic region close to the plasma membrane are illuminated and visualized using the evanescent field. In this way, TIRFM eliminates potential fluorescence background from much of the cytosolic region that may obscure the events close to the cell surface. In spite of the advantages, like most imaging tools TIRFM is limited to the investigation of a small number of cells due to the small size of the illumination and imaging frames. Although useful biological information can be generated using the technique, whether such data are representative of the large cell population remains questionable, especially when heterogeneous cell populations such as those obtained from primary materials are involved. On the other hand, conventional flow cytometry does not differentiate fluorescence emission from different subcellular locations of a cell [8]. High throughput techniques that can provide information on the cell surface features for a cell population with single cell resolution have been lacking.

Here we report a microfluidics-based tool, referred to as total internal reflection fluorescence flow cytometry (TIRF-FC), to examine cells in a flow for their surface features with a throughput ~100-150 cells/s. A microfluidic device with an elastomeric valve creates a constriction to drive flowing cells into an evanescent field which excites only fluorescent species in the region of the cell membrane. Our results show that TIRF-FC is sensitive to the differences in the intracellular localizations of fluorescent molecules. Thousands of cells can be examined with high throughput without requiring adhesion of cells on the substrate. We envision that TIRF-FC can be a powerful tool to study intracellular dynamics and to screen heterogeneous cell populations when biological events at the cell surface are of interest.

## II. EXPERIMENT

**1. Microchip fabrication**. The device was fabricated using multilayer soft lithography [9, 10]. The microscale patterns were first created using a computer-aided design software (FreeHand MX, Macromedia) and then printed out on high-resolution (5080 dpi) transparencies as photomasks in photolithography. The control layer master (photoresist/3 inch silicon wafer) was made using a negative photoresist SU-8 2025 (Microchem) with a thickness of ~25 μm (measured by a Sloan Dektak3 ST profilometer). The fluidic layer master was made using a positive photoresist AZ 9260 (Clariant) with a thickness of ~14 μm. The fluidic layer master was then

[a)]Author to whom correspondence should be addressed; email: changlu@purdue.edu baked at 120 ℃ for 2 min to generate a rounded cro ss section for the channel. The thickness at the center of the rounded fluidic channel was measured to be ~18 um after baking. Both control and fluidic layers of the device were molded with PDMS of the same composition (GE Silicones RTV 615, MG Chemicals, mass ratio of A:B=10:1). The fluidic layer had a thickness of 35 μm (formed by spinning liquid PDMS prepolymer at 4000 rpm for 30s) which left the thickness of the PDMS membrane between the fluidic channel and the control channel ~17 μm. The control layer had a thickness of ~0.5 cm. The two layers were bonded together upon contact after oxidizing the two PDMS surfaces using a plasma cleaner (Harrick). The combined PDMS layers were then bonded to a 45 mm ×50 mm (No.1, Fisher Scientific) glass slip (~170 μm thick) using the same oxidation method. The fluidic channels were conditioned with 1% Pluronic F-127 (Invitrogen) for 1 h before experiment to avoid cell adsorption to the surfaces.

**2. Cell samples.** Syk-deficient DT40 B cells stably expressing SykEGFP (SykEGFP-DT40-Syk$^{-}$) or SykEGFP-NLS (SykEGFP-NLS-DT40-Syk$^{-}$) were produced as described previously [11, 12]. SykEGFP-NLS contains the nuclear localization signal (NLS) from simian virus 40 large T antigen at the C-terminus. Both DT40 cell lines were cultured for at least 15 passages in complete medium (RPMI 1640 supplemented with 10% heat-inactivated fetal calf serum, 1% chicken serum, 50 μM 2-mercaptoethanol, 1 mM sodium pyruvate, 100 IU/ml penicillin G, and 100 μg/ml streptomycin) before experiments in the TIRF-FC devices. Cells were harvested and suspended in phosphate-buffered saline (PBS). To label cells with $DiOC_{18}(3)$ (Invitrogen), $5\times10^6$ DT40 cells in 1 ml were mixed with $DiOC_{18}(3)$ at final concentrations of 0.5, 1, 2, and 4 μM and incubated for 10 min. All cell samples were centrifuged at 300 x *g* for 5 min, washed twice in PBS and re-suspended at a final cell density of $10^7$ cells/ml in PBS before the experiment. Chinese hamster ovary cells (CHO-K1) was cultured according to the protocol applied in our previous study [10]. The CHO cell density was $10^7$ cells/ml in PBS before the experiment.

**3. Optical setup.** An air-cooled 100mW argon ion laser (Spectra-Physics) was applied as the light source for laser-induced fluorescence. The laser beam at 488 nm was spectrally selected by a prism (Thorlabs) and filtered by a 10LF10-488 bandpass filter (Newport) before its intensity was adjusted by neutral density filters (Newport). The laser beam was expanded 5 times in diameter by lens L1 (f =15 mm, Thorlabs) and L2 (f =75 mm, Thorlabs) before it was focused by a lens L3 (f = 400 mm, Thorlabs) and entered the laser port B of an invert fluorescence microscope (IX-71, Olympus). The laser was then filtered by a dichroic beamsplitter (505DCLP, Chroma Technology) and focused on the back focal plane (BFP) of a high numerical aperture TIRF objective (PlanApo, oil, 60X, NA=1.45, Olympus). The specimens in the microfluidic channel were illuminated by paraxial rays with Koehler illumination. The diameter of the circular illuminated area was approximately 90 μm. The intensity of the 488 nm light that was incident on the coverslip was ~2.1 mW. We were able to switch the mode of illumination between widefield epi-fluorescence and TIRF by either shifting the position of lens L3 vertically or pivoting M1. The fluorescence light from specimen was collected by the same objective. After passing through the dichroic filter and emitter (D535/40 emission filter, Chroma Technology), the emitted light was collected either by a photomultiplier tube (R9220, Hamamatsu) biased at 730 V or a CCD camera (ORCA-285, Hamamatsu). The 28 mm-diameter side-on photomultiplier tube was able to collect the light from the entire illuminated area when the flow cytometry data were taken.

**4. Microchip operation.** The microfluidic device was mounted on an inverted fluorescence microscope with an oil immersion 60X TIRFM objective. The 3 inlets were connected with two syringe pumps (PHD infusion pump, Harvard Apparatus) through plastic tubing. The volumetric flow rates were set at 200 µl/h for the sample inlet and 400 µl/h for each of two side inlets to obtain hydrodynamic focusing. The width of the sample stream under the valve was around 20 µm (the rounded shape of the cross section of the fluidic channel helped decrease the sample stream width). The control channel was filled with deionized water to prevent air from leaking into the fluidic channel. The microfluidic elastomeric valve was actuated by pressure provided by a nitrogen cylinder through a fast-response solenoid valve (ASCO Scientific) (Figure 1). The solenoid valve was controlled by a valve control circuit (described in detail below) which quickly open-closed the valve in case of adsorption of cells to prevent clogging.

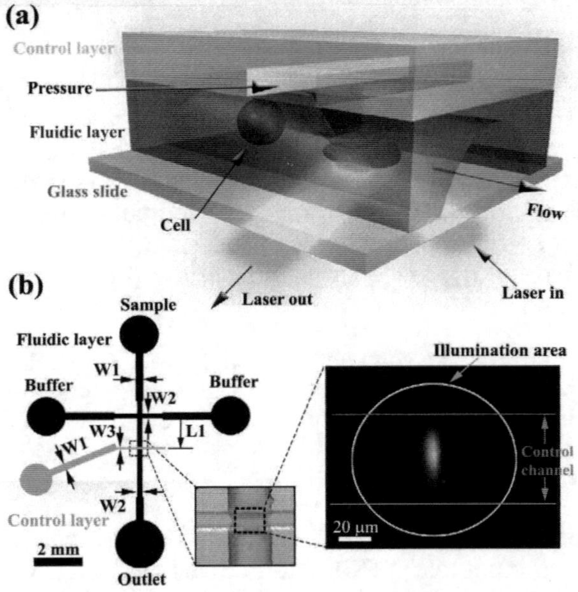


**Figure 1.** (a) A schematic illustration of the microfluidic device for screening fluorescent species at the cell surface (not to scale). The partially-closed elastomeric valve renders a flowing cell in contact with the glass cover slip where there is an evanescent field created by an incident laser beam (488 nm). (b) The layout of the microfluidic chip and the TIRF illumination of flowing cells. The control channel (gray) is on top of the fluidic channel (black) with a polydimethylsiloxane (PDMS) membrane in between. The depth is 25 µm for the control channel. The rounded fluidic channel has depth of ~18 µm at the center. The dimensions are labeled (W1=300 µm, W2=200 µm, W3=50 µm, and L1=1.5 mm). The inset fluorescent image (right) shows the fluorescent trail left by flowing fluorescent cells (SykEGFP-DT40-Syk$^-$).

**5. Valve control.** The elastomeric valve was kept partially closed during the operation of our microfluidic device in order to push flowing cells into contact with the glass surface. However, clogging and accumulation of cells could occur, depending on the cell size and deformability. In order to solve the problem, we designed a circuit to control the valve so that it could actuate (open and then go back to its partially closed state again) when a fluorescence intensity higher than a threshold is detected (usually caused by an adsorbed fluorescent cell). This design greatly alleviated the clogging problem so that we could continuously screen at least ~4000 cells in a run. The valve was actuated for ~10 times during a typical run with ~4000 DT40 B cells flowing through. It was actuated for less than 5 times when CHO cells of a similar sample size were tested (CHO cells appeared to have higher deformability than DT40 B cells).

The fluorescence signal collected by the PMT was processed by a low noise preamplifier (SR570, Standard Research Systems) with the cutoff frequency and the sensitivity set at 10K Hz and 100 µA/V, respectively. The current signal from the preamplifier was transformed to voltage $V_F$ which was processed by the valve control circuit. The valve control circuit was composed of an active low pass filter, a comparator, and a reed relay. The configuration of the circuit is shown in Figure 2a. $V_F$ was firstly filtered and amplified by a circuit with resistors R1/R2, a capacitor and an operational amplifier (LM358P, National

Semiconductor). Only low frequency signal was allowed to pass in order to prevent flowing cells of high fluorescence intensity from triggering the valve actuation. The cutoff frequency is defined as

$$fc = \frac{1}{2\pi R_2 C} \tag{1}$$

where $R_2$ has a resistance of 9820 Ω and C has a capacitance of 10 μF so that *fc* is 1.62 Hz. The processed signal $V_P$ has a gain of 20 determined by

$$\text{gain} = -\frac{R_2}{R_1} \tag{2}$$

where $R_1$ was 490 Ω. By comparing with $V_{ref}$ (3.02 V), the comparator (LM 339N, Motorola) operated the reed relay (R56-1D.5-6D, NTE Electronics) to control the solenoid valve and hence the microfluidic elastomeric valve. The clogging event with a fluorescent cell has a median to high level signal with low frequency, which triggered the valve to open and consequently remove the clogging cell. The valve returned to its original state (partially closed) immediately after removing the clog. The response time for the valve is less than 2 ms. The actuation of the valve causes a negative signal in the fluorescence intensity data (Figure 2b), due to the reflection in the circuit when it responded.

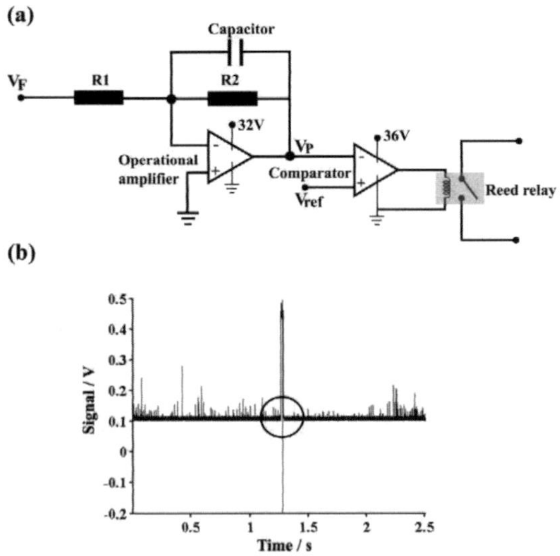


**Figure 2.** (a) The configuration of the valve control circuit. (b) Fluorescence intensity data recorded by LabView revealing one clogging event and the subsequent removal of the clog with the valve actuation (circled in the figure).

**6. Signal processing.** $V_F$ from the preamplifier was input into a PCI data acquisition card (PCI-6254, National Instruments) operated by LabView software (National Instruments). When the device was in operation, the fluorescence intensity recorded in voltage by LabView at $10^6$ Hz showed a series of spikes with each of them corresponding to one cell flowing through the detection window. The data were processed by programs written in MATLAB to determine the width and height of each spike. The width of the spikes ranged from 0.15 to 40 ms. The spikes having dimensions within certain boundaries would be counted as valid data (e.g. the artifact spikes generated by the cell adsorption and the valve actuation would be discarded in this process). Then the data were sorted into histograms of the fluorescence for a cell population. The fluorescence intensity (the height of the spikes) ranging from 1 mV to 10 V was converted to 4 decade logarithmic voltage scale and then 256 scale channels, due to the small sample size of 3000-4000 cells in each histogram. The throughput of cells is 100-150 cells per second through the laser detection spot.

## III. RESULTS AND DISCUSION

As shown in Figure 1, a multilayer microfluidic device with an elastomeric valve was used in the study [9]. Hydrodynamic focusing aligned cells in a single file at the center of the channel in order for them to flow under the elastomeric valve. The valve was partially closed during the operation in order to form a flexible constriction that forced flowing cells to contact the glass substrate due to their deformation. An objective-based TIRF setup was implemented to introduce the evanescent field on the glass surface for cell illumination. A deformed flowing cell had a part of its surface illuminated by the evanescent field and the fluorescent signal was collected by a photomultiplier tube (no fluorescent signal was detected when the valve was not pressurized). Most of cells were able to pass under the partially closed valve due to the deformability of both the PDMS membrane and the cells. We designed a control circuit to actuate the elastomeric valve (open-close) whenever both the fluorescence intensity and the residence time were higher than designated thresholds, in order to remove fluorescent cells that adhered to the valve surface. This practice was necessary and effective for preventing the clogging of the device. Typically our devices were able to continuously screen ~4000 cells with a throughput of ~100-150 cells/s without being clogged.

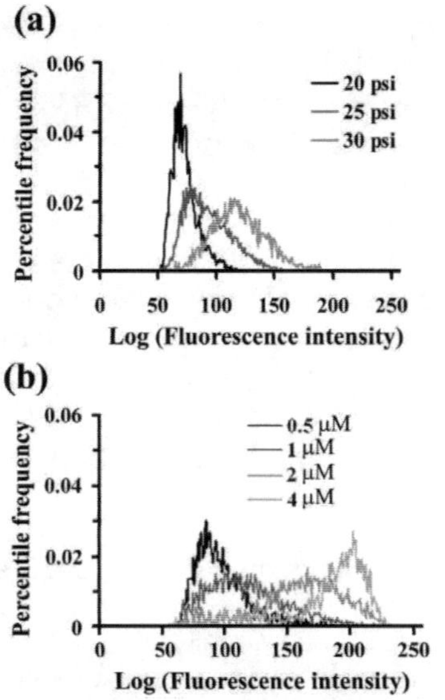


**Figure 3.** The percentile histograms of log (Fluorescence Intensity) from SykEGFP-DT40-Syk⁻ cells and $DIOC_{18}(3)$ stained CHO cells detected using TIRF-FC. All histograms were normalized to eliminate the effect of sample size. The percentile frequency on the y axis is the frequency divided by the total number of observations. 3000-4000 cells were involved in each histogram. (a) Histograms of SykEGFP-DT40-Syk⁻ cells with 20 psi, 25 psi and 30 psi actuation pressure in the elastomeric valve. (b) Histograms of CHO cells stained by 0.5 μM, 1 μM, 2 μM, 4 μM $DiOC_{18}(3)$ for 10 min. The actuation pressure of the valve was maintained at 25 psi in all cases.

We tested the performance of the TIRF-FC using chicken B cells and Chinese hamster ovary (CHO) cells. Chicken B cells with enhanced green fluorescent protein (EGFP) tagged spleen tyrosine kinase (Syk) expression (SykEGFP-DT40-Syk⁻) were used for testing the effects of the actuation pressure of the elastomeric valve on the fluorescence intensity from cells. Syk is an important protein-tyrosine kinase involved in signaling pathways that are critical for cell growth and development[13-15]. SykEGFP locates in the entire cytosol and nucleus in this cell line and the fluorescent signal from the cells was detected by our TIRF-FC apparatus. A histogram of log (fluorescence intensity) could be obtained for a cell population of ~3000 cells. In Figure 3a, we show that higher fluorescence intensity was detected when the actuation pressure of the valve was increased from 20 psi to 30 psi. The mean fluorescence intensity at 20 psi was 71 (in channels), while it increased to 94 at 25 psi and 121 at 30 psi. Higher actuation pressure in the control channel pushed the PDMS membrane (between the control channel and the fluidic channel) closer to the glass substrate and this made cells deform more during the passage under the valve. The more substantial deformation led to larger cell surface area in contact with the glass surface and stronger TIRF signal. We also tested CHO cells (D~14.6 μm with a standard deviation of 2.2 μm) that are larger than chicken B cells (D~10.8 μm with a standard deviation of 1.3 μm). CHO cells were stained with various concentrations of $DiOC_{18}(3)$. As shown in Figure 3b, with 25 psi as the actuation pressure of the valve, the fluorescence histograms taken from CHO cells with various degrees of labeling can be differentiated as the mean fluorescence intensity channel increased gradually from 97 with 0.5 μM $DiOC_{18}(3)$ staining to 195 with 4 μM $DiOC_{18}(3)$ staining.

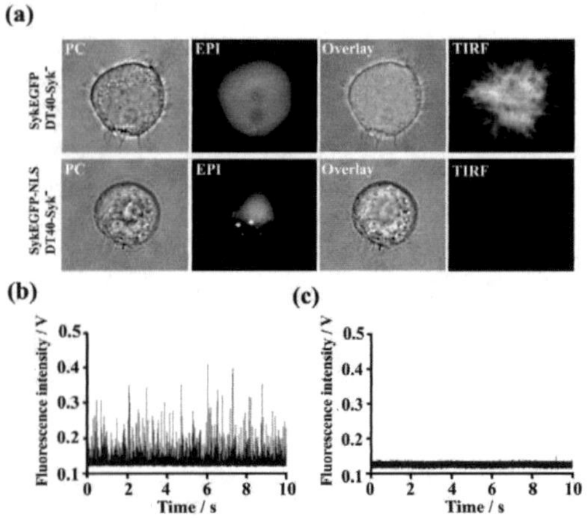


**Figure 4.** (a) Phase contrast (PC), epifluorescent (EPI), overlay and TIRF images of SykEGFP-DT40-Syk$^-$ and SykEGFP-NLS-DT40-Syk$^-$ cells. The images were taken after cells settled down onto a cover slip. (b) Fluorescence intensity recorded when SykEGFP-DT40-Syk$^-$ cells flowed through the detection point for 10 seconds. (c) Fluorescence intensity recorded when SykEGFP-NLS-DT40-Syk$^-$ cells (with the same density as SykEGFP-DT40-Syk$^-$ cells) flowed through the detection point for 10 seconds.

TIRF-FC is capable of determining whether fluorescent molecules are in the proximity of the cellular membrane. We compared chicken B cells transfected with plasmids coding for SykEGFP and SykEGFP-NLS[11]. As shown in Figure 4a, Syk-EGFP is expressed in the entire cytosol and the fluorescent fusion protein can be seen using TIRFM imaging. In contrast, SykEGFP-NLS, which contains a nuclear localizing sequence (NLS) and localizes primarily to the nucleus, cannot be detected using TIRFM. As shown in Figure 4b and 4c, our microfluidic tool was able to differentiate the two cases. Cells carrying SykEGFP-NLS essentially did not generate any fluorescence signal. This confirms that TIRF-FC is effective for observing the differences in the subcellular location of intracellular molecules when the amount of molecules in the periphery of the membrane varies. The dynamics in a protein's subcellular location, or protein translocations, are important events involved in a large number of protein activation processes [16-18]. The recognition of the change in the protein subcellular location is therefore an important trait of TIRF-FC.

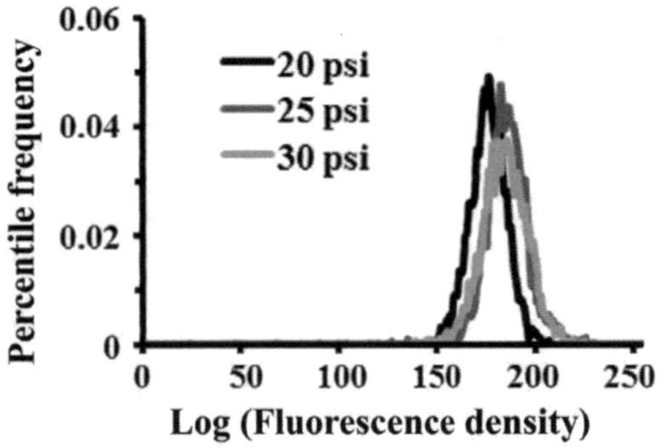


**Figure 5.** The reanalysis of the data from Figure 3(a) by creating the histograms of the area fluorescence density. The area fluorescence density of a cell was estimated by calculating H/W with H and W being the height and the bottom width of a spike generated by the cell, respectively.

The mechanical properties and the operational parameters of the elastomeric valve affect the results and their quantitative analysis. For example, the actuation pressure and the mechanics of the valve affect the degree of the PDMS membrane deformation and consequently these change the contact surface area and the relative position (i.e. whether the cell passes under the center of the valve) of a cell when it flows under the valve. However, such effects can be largely eliminated by proper processing and calibration of the fluorescence data. In Figure 5, we carry out a simple procedure to convert the fluorescence intensity data in Figure 3(a) (taken at different actuation pressures) into the fluorescence density data. For each spike generated by a flowing cell, we extract information on its height (H) and bottom width (W) . We use H×W to estimate the amount of fluorescent molecules detected (we ignore any constant in this analysis). We also assume that the bottom width W of the spike gives a rough measure of the linear dimension of the contact surface. Then we use W×W to estimate the contact surface area for each flowing cell. Based on these assumptions, the area density of the fluorescent species detected can be obtained for each cell by calculating the value of H/W of the spike generated by the cell. Because the three histograms in Figure 3(a) were taken using the same cell type, in Figure 5 we show that the histograms of the fluorescence density are very close to each other despite that they were taken under different actuation pressures. This suggests that with appropriate calibration the operational parameters may have little influence on the area fluorescence density on the cell membrane obtained by TIRF-FC. In principle, the data calibration can be further improved by having more accurate information on the cell size (e.g. by light scattering) and adding correction to compensate the differences in the velocity of the cells while they pass through the detection point (i.e. large cells squeeze through the valve more slowly than small cells and create boost to their detected fluorescence). We believe that proper data processing and analysis will allow TIRF-FC to be a quantitative tool for studying the dynamics at the membrane and its proximity.

## IV. CONCLUSIONS

We demonstrate microfluidics-based TIRF-FC for carrying out high-throughput examination of single cells with evanescent field illumination. The technique provides information on a cell population compared to examining a few cells using imaging. TIRF-FC retains the trait of detecting events at the periphery of the cell surface. Furthermore, our technique does not require adhesion between the cells and the surface. We envision that this technique will provide insight into the dynamics and heterogeneity in a cell population when biological processes involving events close to the cell membrane occur. Future development of the tool will benefit from adding parameters of detection (e.g. light scattering) and improved protocols for calibrating/analyzing the data.


## ACKNOWLEDGEMENT

We thank Wallace H. Coulter Foundation, NSF CBET 0747105, and NIH CA37372 for the financial support of this research. We thank Dr. Christopher J. Staiger for allowing us to use his equipment to carry out some initial testing.


## REFERENCES


1. Axelrod, D. (2001). Total internal reflection fluorescence microscopy in cell biology. *Traffic* **2**, pp. 764-774.
2. Steyer, J. A. & Almers, W. (2001). A real-time view of life within 100 nm of the plasma membrane. *Nat. Rev. Mol. Cell Biol.* **2**, pp. 268-75.
3. Seisenberger, G., Ried, M. U., Endress, T., Buning, H., Hallek, M. & Brauchle, C. (2001). Real-time single-molecule imaging of the infection pathway of an adeno-associated virus. *Science* **294**, pp. 1929-1932.
4. Stephens, D. J. & Allan, V. J. (2003). Light microscopy techniques for live cell Imaging. *Science* **300**, pp. 82-86.
5. Duncan, R. R., Greaves, J., Wiegand, U. K., Matskovich, I., Bodammer, G., Apps, D. K., Shipston, M. J. & Chow, R. H. (2003). Functional and spatial segregation of secretory vesicle pools according to vesicle age. *Nature* **422**, pp. 176-180.
6. Jones, J. T., Myers, J. W., Ferrell, J. E. & Meyer, T. (2004). Probing the precision of the mitotic clock with a live-cell fluorescent biosensor. *Nat. Biotechnol.* **22**, pp. 306-312.

7. Chronis, N. & Lee, L. P. (2004). Total internal reflection-based biochip utilizing a polymer-filled cavity with a micromirror sidewall. *Lab Chip* **4**, pp. 125-130.
8. Wang, J., Bao, N., Paris, L. L., Wang, H. Y., Geahlen, R. L. & Lu, C. (2008). Detection of kinase translocation using microfluidic electroporative flow cytometry. *Anal. Chem.* **80**, pp. 1087-93.
9. Unger, M. A., Chou, H. P., Thorsen, T., Scherer, A. & Quake, S. R. (2000). Monolithic microfabricated valves and pumps by multilayer soft lithography. *Science* **288**, 113-116.
10. Wang, J., Stine, M. J. & Lu, C. (2007). Microfluidic cell electroporation using a mechanical valve. *Anal. Chem.* **79**, pp. 9584-7.
11. Zhou, F., Hu, J. J., Ma, H. Y., Harrison, M. L. & Geahlen, R. L. (2006). Nucleocytoplasmic trafficking of the syk protein tyrosine kinase. *Mol. Cell. Biol.* **26**, pp. 3478-3491.
12. Ma, H., Yankee, T. M., Hu, J. J., Asai, D. J., Harrison, M. L. & Geahlen, R. L. (2001). Visualization of Syk-antigen receptor interactions using green fluorescent protein: Differential roles for Syk and Lyn in the regulation of receptor capping and internalization. *J. Immunol.* **166**, pp. 1507-1516.
13. Zioncheck, T. F., Harrison, M. L. & Geahlen, R. L. (1986). Purification and Characterization of a Protein-Tyrosine Kinase from Bovine Thymus. *J. Biol.Chem.* **261**, pp. 5637-5643.
14. Zioncheck, T. F., Harrison, M. L., Isaacson, C. C. & Geahlen, R. L. (1988). Generation of an Active Protein-Tyrosine Kinase from Lymphocytes by Proteolysis. *J. Biol. Chem.* **263**, pp. 19195-19202.
15. Coopman, P. J. P., Do, M. T. H., Barth, M., Bowden, E. T., Hayes, A. J., Basyuk, E., Blancato, J. K., Vezza, P. R., McLeskey, S. W., Mangeat, P. H. & Mueller, S. C. (2000). The Syk tyrosine kinase suppresses malignant growth of human breast cancer cells. *Nature* **406**, pp. 742-747.
16. Hsueh, Y. P., Wang, T. F., Yang, F. C. & Sheng, M. (2000). Nuclear translocation and transcription regulation by the membrane-associated guanylate kinase CASK/LIN-2. *Nature* **404**, pp. 298-302.
17. Nelson, D. E., Ihekwaba, A. E. C., Elliott, M., Johnson, J. R., Gibney, C. A., Foreman, B. E., Nelson, G., See, V., Horton, C. A., Spiller, D. G., Edwards, S. W., McDowell, H. P., Unitt, J. F., Sullivan, E., Grimley, R., Benson, N., Broomhead, D., Kell, D. B. & White, M. R. H. (2004). Oscillations in NF-kappa B signaling control the dynamics of gene expression. *Science* **306**, pp. 704-708.
18. Matsuzawa, A., Tseng, P. H., Vallabhapurapu, S., Luo, J. L., Zhang, W. Z., Wang, H. P., Vignali, D. A. A., Gallagher, E. & Karin, M. (2008). Essential cytoplasmic translocation of a cytokine receptor-assembled signaling complex. *Science* **321**, pp. 663-668.

# Environmental Stimuli towards Oxygen Gradient - A Fluidic Device with An Integrated Array of Oxygen Sensors

**Satya Gowthami Achanta**, Department of Chemical & Biological Engineering, Missouri University of Science and Technology, Rolla, MO

David B. Henthorn, Department of Chemical & Biological Engineering, Missouri University of Science and Technology, Rolla, MO

Chang-Soo Kim, Department of Electrical & Computer Engineering, Missouri University of Science and Technology, Rolla, MO

Cell or tissue culture mainly depends on different types of gradients of nutrients, pH, etc. For instance, an atmosphere of dissolved gases gradient such as O2, CO2 being created near the environment surrounding the mammalian tissues helps in the study of cell differentiation and cell growth. Our study focuses on integrating an array of commercial oxygen sensors with a fluidic device which creates an oxygen gradient. Oxygen has been selected because of its wide importance in cellular functions. PDMS was used as the main structural material for the rectangular fluidic pattern because of its ease of fabrication and transparency for easy observation of cell growth. Thin PDMS sheets were used as membranes to create the required oxygen gradient in the culture medium. The culture medium used was agar gel in which all the salts and nutrients required for cell growth were present in optimum amount. Commercial optical oxygen sensors were embedded in the system which measures nutrient concentration at different points in the agar gel and surrounding environment.

See more of this Session: Biomems and Microfluidics: Novel Applications
See more of this Group/Topical: Topical 3: 2009 Annual Meeting of the American Electrophoresis Society (AES)

# Cell Focusing in a Serpentine Microchannel with Insulative Dielectrophoresis

**Christopher Church**, Mechanical Engineering, Clemson University, Clemson, SC

Gaoyan Wang, Biological Sciences, Clemson University, Clemson, SC

Jeremy Tzuen-Rong Tzeng, Biological Sciences, Clemson University, Clemson, SC

Xiangchun Xuan, Mechanical Engineering, Clemson University, Clemson, SC

We present the electrokinetic focusing of yeast cells in a serpentine microchannel using DC-biased AC electric fields. This focusing results from the cross-stream dielectrophoretic motion of cells induced within the channel turns. The effects of electric field, buffer concentration, and cell concentration etc. on the focusing performance are examined.

See more of this Session: Biomems and Microfluidics: Novel Applications
See more of this Group/Topical: Topical 3: 2009 Annual Meeting of the American Electrophoresis Society (AES)

# Mass Spectrometry of Biomolecules Using AC Electrospray Ionization

**Nishant Chetwani**, Chemical and Biomolecular Engineering, University of Notre Dame, Notre Dame, IN

Hsueh-Chia Chang, Department of Chemical and Biomolecular Engineering, University of Notre Dame, Notre Dame, IN

David Go, Aerospace and Mechanical Engineering, University of Notre Dame, Notre Dame, IN

Since John Fenn's pioneering work on proteomic mass spectrometry, electrospray ionization (ESI) has become a popular "soft" ionization technique for the ionization of large biomolecules such as proteins without fragmenting them. This method utilizes a sufficiently high electric field to deform the liquid source into a so-called Taylor cone, and a charged liquid jet emerges from the tip of the cone. This jet further undergoes Rayleigh fission to produce sub-micron droplets that further cascade into next generation of charged drops and eventually lead to the molecular ion of the species that is suitable for detection by mass spectrometry.

In the present work, we demonstrate the implementation of a high frequency AC electrospray as an ionization technique for mass spectrometry. Unlike the familiar DC Taylor cone that has a steady half cone angle of ~ $49^{o}$ , an AC cone has a much sharper tip, with a half cone angle ~ $11^{o}$ and shows a continuous temporal growth. In our earlier work, we have shown that the application of a suitable high frequency AC electric field, where the period of the AC signal $\omega^{-1}$ is less than the Maxwell relaxation time scale of the liquid jet ( $\sigma$ and $\varepsilon$ are the conductivity and permittivity of liquid, respectively), preferentially entrains low mobility ions towards the tip of the cone. The Coulombic repulsion between these ions leads to the formation of a slender ac cone and is responsible for its continuous axial growth. Hence, whenever the anions are the low mobility ions, AC electrospray serves to generate an efficient ionization means for negative mode mass spectrometry. This is in direct contrast to its DC counterpart, which tends to undergo electron discharge in negative mode and often requires the use of electron scavenger gases such as oxygen and $SF_6$ to capture the excessive electrons. Moreover, the charge per unit volume within the droplets ejected by an AC electrospray is shown to be an order of magnitude less than that of a DC electrospray, which goes on to suggest that an AC electrospray may serve as a "softer" ionization technique. Here, we present results from negative mode mass spectrometry on AC electrospray of solvents, oligonucleotides and neuropeptides and provide a direct comparison of this novel ionization technique with DC electrospray Ionization.

See more of this Session: Biomems and Microfluidics: Novel Applications
See more of this Group/Topical: Topical 3: 2009 Annual Meeting of the American Electrophoresis

Society (AES)

# AUTHOR INDEX

# AUTHOR INDEX

# AUTHOR INDEX